Sarah Maria Burnham

Precious stones in Nature, Art, and Literature

Sarah Maria Burnham

Precious stones in Nature, Art, and Literature

ISBN/EAN: 9783337024857

Printed in Europe, USA, Canada, Australia, Japan

Cover: Foto ©berggeist007 / pixelio.de

More available books at **www.hansebooks.com**

PRECIOUS STONES

IN

NATURE, ART, AND LITERATURE

BY

S. M. BURNHAM

AUTHOR OF "LIMESTONES AND MARBLES: THEIR HISTORY AND USES"

Boston

BRADLEE WHIDDEN

1886

CONTENTS.

PREFACE.

In the preparation of the work on "Precious Stones," use has been made of all the assistance within reach of the author obtained from writers expressly discoursing upon this topic, from general literature, and from works exclusively scientific. The subject covers a wide field, and is one about which, on some points, there is considerable difference of opinion; therefore, the difficulty of arriving at the truth is much greater than where there is more concurrence of views, and more agreement in the statements of what are claimed to be historical facts.

Very few American works on precious stones exist, and those that have been published are generally limited in the number of species described; while nothing, or only very meagre accounts have been given to American gems by foreign authors. The most complete list of native gems is by Mr. George F. Kunz, mineralogist and gem-expert, employed by Messrs. Tiffany and Company, New York, published in the United States Geological Surveys, under the superintendence of Mr. Albert Williams, Jr. The author very gratefully acknowledges the assistance generously offered by Mr. Kunz, who examined parts of the manuscript, and suggested several improvements, more especially in the scientific and practical portions of the work.

It was the aim of the author to present some facts in reference to the resources of our own as well as of other countries of the globe, in decorative stones for architectural purposes, in a work published in 1883, on "Limestones and Marbles." The present volume is intended as a sort of supplement, covering the same ground, and illustrating the use of precious stones in decoration, more especially as personal ornaments. The gem minerals are, in a certain sense, complemental to the architectural decorative stones, and the crowning glory of nature's handiwork, the rarest of all her material productions, and those invested with the greatest fascination, either as objects of careful study or as treasures to be won at great sacrifice.

It is not easy, nor, perhaps, desirable, to give a list of all the writers from whom assistance, either directly or indirectly, has been obtained, since they are scattered through various departments of knowledge, but many of them have been referred to in the text, and if a thought has been borrowed without specifying the definite source whence it was taken, it has usually been because it could be traced to several authors, showing that either the idea was native to all, or that all had obtained it from the same source.

Doubtless there are many errors which have been overlooked by the author, or which have been received as truth; but it is hoped that facts have been presented in most instances and made instructive to the general reader.

PRECIOUS STONES.

CHAPTER I.

ORIGIN, PROPERTIES, CLASSIFICATION, LOCALITIES, IMITATIONS, AND ANTIQUITY OF PRECIOUS STONES.

Origin. — A desire to penetrate the hidden mysteries of nature's operations is innate in man, and has led to some of the grandest and the most useful achievements of the human mind. This longing to become acquainted with her laws and to account for her phenomena stimulated the activity of ancient thought as it now incites modern investigation, and has given birth to many of the innumerable theories that have always marked the progress of science.

The various speculations in regard to the origin of precious stones afford some curious illustrations of the mental peculiarities of different nations as well as individuals; as, for example, the Greeks, with their poetical and religious biases, referred them to the direct agency of their divinities, or to some of the forces of nature personified and invested with mysterious powers. The youth who rocked the cradle of the infant Jupiter on the Island of Crete was transformed into the adamas, and here we have the origin of the diamond. A beautiful nymph beloved by Bacchus was changed into the amethyst, representing the color of this god's favorite beverage. The sources of amber were numerous: drops of perspiration exhaled by the goddess Ge, — the Earth, — the tears shed by the sisters of the ill-fated

Phaeton, the tears shed by the sisters of Meleager, the tears for Æsculapius, the tears of certain sea-birds, to which allusion is made by the poet in the lines : —

> "Around thee shall glisten the loveliest amber
> That ever the sorrowing sea-bird hath wept."

Amber is certainly a most pathetic gem, since so many tears were shed at its birth ; but it had also a more material source — honey melted by the sun and congealed by falling into the sea. Lapis-lazuli sprung from the agonizing cry of an Indian giant ; the emerald originated in the fire-fly ; and other equally fantastic notions constituted the popular belief in regard to the origin of precious stones, though some of the ancient philosophers were disposed to account for their existence on less superstitious grounds. They were supposed by Plato to be the result of fermentation originating in the stars, while the diamond, which has always been an exceptional gem, was the kernel of auriferous matter condensed into a transparent mass. Theophrastus, nearly twenty-four centuries ago, discarding the general belief in the supernatural origin of mineral species, thought all rocks and metals originated from water and earth, water being the base of metals, earth of stones, both common and precious.

Later Theories. — Modern scientists, who are as much inclined to speculations as their predecessors, have their extraordinary and conflicting systems upon this fruitful theme. There are those who maintain that precious stones are the result of aqueous solution, others that they were the product of hot vapors, while a third school believe they were formed through metamorphism by segregation from older rocks ; but how the primary rocks came into being, is a question which naturally arises for solution. Robert Boyle, of the seventeenth century, believed all precious stones were originally formed of limpid water, and that their color and other essential properties were derived from their metallic spirit. Sir John Hill, nearly a century later, adopted the opinion that they were formed by the concretion of matter from cohesion or by some kind of percolation, and that the difference of their constituents and the manner of coalescence were the causes of their various qualities, as smoothness, density, transparency, etc. He further maintained that their constituent matter was a pellucid, crystalline substance of different degrees of hardness, and had it been in a perfectly pure state, all precious stones would have been without color.

Haüy, the father of modern mineralogy, says most crystals were formed in water where the constituents, at first separated and suspended, were brought together by force of mutual attraction; that is, the particles diffused and floating were brought together by the attraction of cohesion and precipitated, when they formed a stratum pure and homogeneous. This constitutes the aqueous theory, which has its opponents.

As most precious stones are transparent or translucent, the inference has been drawn that their constituents must have been in the condition of gases or liquids — an opinion sustained by the discoveries of the microscope, which reveal the fact that in many different species, water or some other fluid is enclosed in cavities, often so extremely minute that several millions occur in a cubic inch. These little cells appear luminous by reflected light, which gives brilliancy to the gem; but if the light be transmitted, they present a dark outline. Some of these porous crystals not unfrequently burst and fly to pieces by the application of strong heat, in consequence of the expansion of the enclosed fluid.

Water often forms one of the constituents of rocks, but it is in a different state from that found in cavities, which makes no part of their substance. Sir David Brewster believed every mineral enclosing water was of aqueous origin, but Mr. Morris says we are not to suppose the presence of water essential to the formation of crystals, since they are also produced by igneous fusion, when the cavities are filled with a substance resembling glass, as seen in augite from Vesuvius. Sometimes the matter enclosed is crystallized, when the pores are called stone cavities, and at other times the cells are filled with gas. The fluid cavities of zeolites — "boiling stones" — seem to indicate that they were deposited in heated waters.

Minerals found in a conglomerate of Mount Somma enclose all the different kinds of cavities, showing, says this writer, they were made by the combined action of water and igneous fusion. He thinks the minerals of Mt. Vesuvius were formed at a dull heat of 335° Centigrade, under a pressure of, probably, two thousand feet, and in the presence of water holding alkaline salts in solution, different gases, and vapors.

Mr. Church is of the opinion that the natural process of forming precious stones was by water, great pressure, and long time; while another writer on the subject divides them, in reference to their origin, into two classes: those formed by direct fusion, the igneous method; and those by water, the aqueous method. The hypothesis may be confidently assumed that the elementary constituents of precious stones existed in a state to move freely among themselves, or their homogeneous character could not have been secured. This condition, it is maintained, could have been obtained by fusion, by disintegration, or by reduction to vapor.

These various theories may all comprise some truth, but they leave the subject open to further investigation.

Properties. — There is a distinction between precious stones and gems in a strictly scientific sense. The name "precious" applied to a mineral refers to only a few species, generally distinguished by superior transparency, lustre, color, hardness, and some other characteristics; while "gem" is a term which embraces a wider range, and comprises a larger variety of materials used for personal decoration. In a popular sense, however, precious stone and gem are nearly identical, and include several substances not mineral, and others, which are wanting in some of the qualities considered essential in an ornamental stone of the first class.

Writers on precious stones differ materially in the classification and arrangement of their properties, some of the older mineralogists making color the test of their distributive order, while modern scientists class them according to their chemical constituents, which consist largely of carbon, aluminum, silicon, magnesium, glucinum, zirconium, and iron, with alkalies for solvents. The excellence of precious stones, it has been said, depends not so much upon their composition as upon the complete solution and combination of their constituents.

Their physical properties are color, lustre, hardness, specific gravity, refraction, polarization, fusibility, combustibility, phosphorescence, and crystallization.

Color. — This is one of the most striking and important qualities of ornamental stones, and constitutes their most attractive feature, always excepting colorless diamonds, and some other species of the first rank. It affords, also, some of the most interesting phenomena connected with these marvels of creation. They may exhibit only one color, and are, therefore, monochroic; they may have more than one, when they are called pleochroic; they may be opalescent, or prismatic, and display all the colors of the rainbow; again, they may reflect rays differing from the color of the crystal, when they are said to be fluorescent; and chatoyant, when they emit a changeable, wavy light.

The colors found in precious stones are the most brilliant in nature, and resemble more closely the hues of the solar spectrum than those of any other material substance. The cause of these different colors has given rise to considerable speculation, in which different opinions have been advanced by different theorists, some having thought that light, and crystalline or molecular arrangement, had an influence in producing them; while others have maintained a different view. It is generally admitted that the coloring matter consists of various metallic oxides; but of the nature of these oxides there is not the same uniformity of opinion. Were the crystals perfectly free from foreign substances, they would all be without color, as is the case with some gems; therefore, the delightful charm arising from the beauty and variety of their hues would have been wanting.

It is conceded that the tone and character of color in precious stones depend upon the nature and quantity of the extraneous substance, combined with the original constituents. This matter thus introduced — iron for instance, which forms one of the most general coloring agents — is not in its elemental state, but is united with oxygen, in different degrees, so as to produce different hues, by changing its density. For instance, one amount in a molecule will give red rays; but by changing the quantity of oxygen the result will be yellow rays, while another combination will afford green rays, etc. Sir John Hill advances some interesting theories on the subject, which later writers have not fully adopted. He says lead produces yellow tints; iron, red; tin, black; copper, green or blue, depending upon the nature of the solvent. If an acid, it will be green; if an alkali, it will be blue. When lead becomes the coloring agent, the crystal is a topaz; when lead is combined with iron, the union forms a hyacinth; but when iron alone is present, a garnet or some other red gem is the result. Ruby owes its hue to gold; but if the crystal is colored by an acid copper, it becomes an emerald; and if by an alkali solvent, a sapphire. This writer advances the opinion that copper, being affected by every kind of solvent, produces an almost infinite variety of beautiful colors, and is probably the base of the coloring matter of more gems than any other, or all other substances combined. He also believed that the coloring agent in precious stones had an influence on the form of the crystals: the cube being the result of lead; the rhombohedron, of iron; the four-sided pyramid, of tin; and other geometrical figures, the effect of other different coloring materials.

In opposition to these views, Haüy says the principal coloring agent in precious stones is iron, with few exceptions, the spinel and Peruvian emerald being colored by chrome, and the chrysoberyl by nickel. In the oriental or precious corundum, iron combined with different quantities of oxygen causes nearly all the colors of the solar spectrum, as seen in the ruby, sapphire, emerald, topaz, and amethyst varieties of this species.

Arranged according to color, the *white*, or, more properly, colorless, species, include the diamond, sapphire, topaz, zircon or jargoon, beryl, phenakite, rock-crystal, and some others, though all or nearly all of these have their colored varieties.

The best known *red* gems are the ruby, of many shades, the spinel (displaying scarlet, flame, and aurora tints, sometimes approaching crimson and violet), the garnet, and the tourmaline.

Orange and *yellow* stones are found with the zircon (which is sometimes compared to transparent gold), essonite, Brazilian topaz, sapphire, chrysoberyl, and beryl.

For *green* gems, we have the emerald, chrysoprase, tourmaline, peridot, garnet of the Urals, aquamarine, and beryl.

Blue stones comprise the sapphire, spinel, iolite, lapis-lazuli, and indicolite (a variety of tourmaline); while for *purple*, the amethyst and the almandine garnet afford examples.

The color of some precious stones, when looked *at*, is different from that seen when looking *through* the crystal; that is, their reflected rays are not like those transmitted, as is the case with the tourmaline and the sapphire d'eau, or iolite.

Dichroism (the quality of exhibiting two colors) and *pleochroism* (the quality of exhibiting more than two) are, it is supposed, due to the refraction and polarization of light; and it is only in double-refracting crystals that these properties inhere. The different colors displayed by these minerals depend upon the direction in which they are viewed.

The tourmaline affords the best illustration of this remarkable property, presenting, as it does, red, yellow, green, and blue, with some other tints, in the same crystal; the emerald and the ruby exhibit the same phenomenon, only in a less degree. A blue sapphire, examined by a dichroscope, affords, besides its ordinary color, a greenish yellow, the topaz a pink and yellow; while andalusite, from Brazil, gives white, green, and pink hues.

Lustre, an important quality in gems, depends upon their structure, texture, and reflecting powers. There are different kinds of lustre, as adamantine, like the diamond; vitreous, like glass; resinous, pearly, and silky. The brilliancy of a precious stone is the result of its lustre, or its power of receiving a polish, and may differ in the same species.

Transparency.—The capacity of transmitting light enhances

the value of decorative stones, as in the diamond, which probably affords the best illustration of this quality; but there is a wide difference in the degree of this power possessed by most gems, which receives different names, according to its strength or feebleness. They are transparent, when objects are distinctly seen through them; translucent, when light passes through, but no objects are seen; and opaque, when no light is transmitted. Some transparent gems become more or less opaque when seen in certain directions. Writers on stones sometimes use the first and second terms indiscriminately, calling a mineral transparent when it is only translucent. The ancients accounted for the lustre and transparency of the diamond by supposing it was congealed water.

Hardness in precious stones is of great importance, since it protects them from injury, renders them capable of a high polish, and fits them for testing this quality in other species of minerals. The property of hardness does not mean the power of resisting crushing weight, since a very hard mineral may be very brittle; nor does it depend upon the tenacity with which the particles cohere, or its infrangibility, since the hardest stones, like the diamond, may be easily broken by a fall or a blow; but it implies the quality of resisting the action of a point, — as of a needle, — or the difficulty of being scratched by any softer substance.

The brilliancy and fire, or play of colors, are, to a certain extent, influenced by the hardness of the substance, though not in all cases, as in the opal. The diamond will not yield to any other stone, but will scratch all others; hence, it is ranked as the hardest gem. The sapphire will resist quartz, proving the latter to be the softer. The scale of hardness established by Mohs ranges from 1, the softest, to 10, the hardest, a place assigned to the diamond alone. Some of the best known precious stones are arranged by mineralogists in the following order of hardness:—

Diamond	10	Emerald	7.8	Jade	6.5
Sapphire	9	Zircon	7.8	Peridot	6.3
Ruby	8.8	Tourmaline	7.5	Moonstone	6.3
Chrysoberyl	8.5	Phenakite	7.5	Turquoise	6
Spinel	8	Almandine	7.3	Opal	6
Topaz	8	Iolite	7.3	Lapis-lazuli	5.2
Aquamarine	8	Amethyst	7	Callainite	4

For a more extensive list, see Table of Hardness and Specific Gravity.

It will be seen that different species have sometimes the same degree of hardness, and the question arises, how are they to be distinguished from one another. In reply to this query, it may be said that hardness, though an important test, is not the only one, nor is it always the best one to be used, especially when a gem might be injured in its application. A safer and perhaps a more satisfactory criterion is

Specific gravity, which is the weight of a body compared with the weight of an equal bulk of water,—that is, it expresses how many times heavier it is than water; or it may be defined as the ratio between the weight of the substance and that of an equal volume of some other substance taken as a standard. All solid bodies sink or float in a liquid, according as their specific gravity is greater or less than that of the liquid. Therefore, taking distilled water as 1, all bodies with a specific gravity greater than 1 will sink if plunged into it. The rules for obtaining specific gravity vary. One method is to divide the weight of the body—a precious stone, for instance—in air by its loss of weight in water, and the quotient will be its specific gravity. If a gem weighs four grains in air, and three in water, it is evident it has displaced one grain of water, and has a specific gravity of 4. Water is generally used in the application of this test, but a substance known as "Sonstadt's Solution," a double iodide of potash and mercury, with a specific gravity of 3, is sometimes substituted for water in the case of precious stones.

Archimedes (287 B. C.) is said to have been the first to discover and apply the test of specific gravity, by which he detected a fraud perpetrated by a jeweller of his time. Hiero, King of Syracuse, ordered a crown to be made of pure gold; but, suspecting the goldsmith had used alloy in the work, he submitted the diadem to be tested by this eminent mathematician, who not only found that the gold had been debased, but also the exact amount of alloy mixed with it. This method of testing the precious metals and the precious stones was not only used by the Greeks, but also by the people of India, at a very early period in the history of the jeweller's art.

Fusibility and Combustibility.—These properties are not identical, since some precious stones are combustible but not fusible, as is the case with the diamond. The effect of heat on different species of gem-minerals varies: with some it changes their color; with others, it causes them to form globules, to swell and decrepitate, or to become enamel; others acquire the property of phosphorescence when subjected to heat; while some are reduced to powder by the same agency. But nearly all precious stones are infusible, unless combined with foreign substances, as soda or borax. The garnet is an exception. Chemicals of a certain kind will affect some gems, while others resist their influence altogether. Examples of the latter are afforded by the diamond, corundum, and spinel; but the turquoise, garnet, chrysolite, and tourmaline are affected by acids, and the opal by potash.

Phosphorescence.—Certain substances, after exposure to the rays of the sun, remain luminous in the dark for a limited time — a quality denominated phosphorescence. The diamond, in some of its varieties, naturally possesses this attribute to a certain extent.

Electricity. — The property of attracting or repelling certain substances inheres in some bodies, while in others, naturally non-electric, it may be excited by heat, friction, or pressure. Precious stones are more or less electrical either positively or negatively, while some species are positive at one end of the crystal and negative at the other, as the tourmaline.

Another important quality in precious stones for the purposes of jewelry is that of splitting in definite directions, which is called *cleavage.*

Isomorphism is a term applied to crystallized compounds formed of substances differing essentially in their nature, but appearing to be identical.

Crystallization is the property which certain substances possess of solidifying in regular shapes. The name crystal, "ice," was given to quartz by the ancients, from the belief that this mineral was solidified water. Sometimes the term is inaccurately applied to flint glass, which is not crystalline in nature. When bodies cool in solid mass, as in the case of some precious stones, they are said to be *amorphous.*

"The process of crystallization," says Professor Cook, "is one of the most striking phenomena in the whole range of experimental science. Beautiful, symmetrical forms shape themselves in an instant, out of a liquid mass, revealing an architectural power, in what we call lifeless matter, whose existence and controlling influence but few have probably realized."

The substance at the time of crystallization is thought to have been in a state of fusion, gas, vapor, or solution. In the formation of crystals, a different law predominates from that which controls organized beings; that is, they grow externally, and are destitute of any internal organization whatever, corresponding to that found in the vegetable and animal kingdoms. All crystals, unless interfered with, have the power of assuming a definite form, which they retain as a distinctive characteristic; therefore crystallization becomes an important test in determining the kind of precious stone where the resemblance in other qualities is striking, as between a diamond and rock-crystal. They are all classed with one or other of the six systems recognized by modern mineralogists. In their manner of growth, crystals adhering to the faces of rocks have their longest axis at right angles to them, or they may be said to be placed in relation to the rocks as trees are to the soil, as may be seen in some geodes, where they are displayed to advantage.

Optical Properties. — These include *refraction* and *polarization* of light. When a ray of light falls obliquely on the surface of a transparent body, it is refracted, or bent from its original course. Refraction is either single or double. A crystal is said to possess single refraction when only one object is seen through it; but if the rays of light are separated so as to pass in different directions, thus presenting two images, the crystal is called double-refracting, and affords one of the most curious phenomena in nature. This property varies greatly in different gems: those belonging to the monometric system, like the diamond, are single refractors; those of other systems, like the ruby, quartz, and many others, are double refractors; the topaz and the tourmaline are particularly distinguished for double refraction. *Dispersion* is the property of a refracted ray to separate into its constituent colors, and produces the prismatic effects so delightful in gem-stones.

Polarization of Light. — It is thought that each luminous molecule has two poles, analogous to the poles of the magnet, and that a beam of light reflected at a certain angle will be again reflected if the two plates are parallel to each other, but not if they are perpendicular to each other; this beam is said to be polarized. In some double-refracting crystals, the two opposite polarized beams of light are of different colors.

Classification. — Hardly any two writers concur in the same system of classification, and perhaps in no other department of scientific knowledge have there been so many arbitrary arrangements of a subject as in that of precious stones. This may have arisen from a difference of opinion in regard to their true character, and ignorance of their chemical constituents and the laws which govern their crystalline forms. It has been said that no strictly scientific classification of gems is possible; but the nearest approach to it can be reached through their chemical properties, and habits of crystallization. Grouping them according to color, important for some purposes, is the most striking method, and the one most frequently employed by the ancients, but it is entirely misleading in regard to their real nature. The practical artist classes them in reference to color, transparency, brilliancy, and some other attributes; the dealer ranks them in the order of their commercial value and the varying moods of fashion. Another arrangement is to call all those of superior excellence "oriental," though they may never have been brought from the east, and those of inferior quality "occidental," without regard to the place of their origin.

As it is impossible to classify precious stones in any regular system depending upon their beauty, color, transparency, or any other external quality, since the same species often presents a great diversity in these attributes, that method of grouping them according to their chemical composition is probably the best which has yet been employed. In examining a gem-mineral to ascertain the species to which it belongs, it is necessary to establish the nature of its elements and the form of its crystals, but one test alone is not sufficient for this purpose.

Localities. — There is no law, it has been observed, regulating the geographical distribution of mineral species, as is the case with plants and animals, hence climate has little or no influence upon their development, yet it is a fact that the richest colored gems are found in tropical regions.

They occur in different geological formations, but the most valuable are found in the oldest. Sometimes they are imbedded in a mass of rock, at other times they are near the surface, in diluvial or alluvial soil, gravels, and sands of river-beds where they are seen as river pebbles, and not unfrequently do they appear in derivative rocks, far from their original home.

They are most abundant in warm countries, and from this circumstance it has been thought that volcanic agency may have had some influence in producing them. It would seem that "some peculiar conditions in the laboratory of nature," must have been required for the production of these her choicest gifts. Some of the southern countries of the eastern continent yield the finest and the largest quantities of the most valuable gems, — the ruby, sapphire, topaz, spinel, jacinth, and other colored stones. How can this be accounted for except on the ground that climate has to some extent a controlling effect upon the formation of precious stones, though it cannot be the only influence, since they occur, in some of their species, in nearly every country on the globe.

Although there are many places in the United States where they have occasionally been found, yet it has been stated in the reports on our "Mineral Resources" that there are but two states, Maine and North Carolina, where systematic mining for precious stones has been carried on. Some attempts have recently been made in Colorado, which have resulted in securing good specimens of topaz, phenakite, and amazon-stone, of considerable value; in other instances, native gems have been discovered, not as the fruit of special effort for this object, but incidentally, or in connection with mining for gold or other substances.

North Carolina is probably the richest state in the Union for its gem-minerals, many of which are of the first class. A few specimens of the diamond, of small size but excellent quality, have been discovered in six different counties in this state. The corundum, though abundant in other localities in some of its varieties, affords here gems of the first rank among precious stones, which have been successfully mined through the enterprise of Col. C. W. Jenks. Zircon has appeared in several places, in small, transparent crystals; garnets, agates, malachite, opal affording specimens for gems, spodumene, hiddenite, beryl of rich, deep green, spinel, azurite, amethyst, rose-quartz, sagenite, rutile, and aquamarine, all suitable for ornamental stones, have been obtained from North Carolina. The collection sent to the New Orleans Exposition comprised a beautiful variety of white beryl, and another of a rare shade of yellow, varieties of quartz, fine specimens of hiddenite, emerald, spodumene, ruby, aquamarine, rutile, jasper, Venus-hair-stone, remarkable specimens of quartz inclosures, amazon-stone, citron-topaz, and other kinds of gem-minerals.

California offers a considerable variety of ornamental stones, including the diamond, corundum, opal, garnet, various kinds of the quartz species, malachite, azurite, selenite, and obsidian. A wonderfully clear specimen of quartz with moss-like inclosures afforded a very rare and interesting feature of the mineral department of this state at the Exposition, on account of the remarkably beautiful effect it produced. The Suisun marble, or aragonite, constitutes a very desirable material for some kinds of decorative work, and may rank with the Mexican onyx.

Artificial Gems.—There is a difference between an artificial stone and an imitation; in the latter, there is an entirely different chemical composition, while an artificial gem can be manufactured from the same chemical substances, and with the same physical properties as the natural specimen, and can be made even to excel the genuine production in brilliancy and play of colors. For instance, the corundum may be obtained by a chemical process, with the same form of crystals, and of the same density and hardness as the ruby and the sapphire, while the artificial spinel cannot be distinguished from Nature's work, by the eye. The same is true of other precious stones more especially the compounds of silica.

Artificial rubies are secured by heating alumina for a long time in a platinum vessel with borax, after which they present the same crystalline form, hardness, and dichroism as the real gem. By repeated experiments, chemists have succeeded in making what were supposed to be artificial diamonds, but with them, as with other precious stones originating in the laboratory, they are too minute for practical purposes. No artificial gems are known in commerce.

Imitations have been secured with much greater facility than artificial varieties, and may be produced of any required size. They are generally made of flint glass and lead, colored by certain oxides as cobalt, manganese, nickel, copper, iron, chrome, and some other substances, the composition being called strass, from the name of the inventor, or paste, whence the name "paste jewels." Glass jewels are not a modern invention, for as soon as the secret of making glass was understood, it was employed in imitating precious stones; in the time of Pliny, the principal gem-minerals were frequently imitated; the emerald, being one of the easiest to counterfeit, was oftenest selected for that purpose. Bracelets of black glass found in the ruins of Chaldæa prove how early such imitations were used for ornaments.

An improvement has been made in the quality of the composition used, by which a superior kind and a greater variety of imitations are now obtained. Strass or paste requires the very best glass mixed with quartz, boracic acid, caustic potash, arsenic, and oxide of lead, with different substances for color, as antimony and gold for topaz, oxide of copper or chromium for emerald, and the oxide of cobalt for sapphire. The imitation of opal requires several different constituents, in which bone ashes are added to various chemical substances.

Pearls have been imitated with great success both in ancient and modern times. The more ancient method was by filling glass beads with a pearly varnish, but this process was improved at a later period by a Frenchman, as the result of studying the habits of a certain species of fish, the *Cyprinus alburnus.* He observed that the water in which this fish was dressed, was filled with small silvery particles which were precipitated to the bottom of the vessel, forming a sediment of a beautiful pearly lustre, to which he gave the name *essence d'orient,* or essence of pearl.

This sediment suggested the idea of using it for the production of false pearls, a result secured by using glass beads covered inside with this pearl "essence" and a solution of isinglass, and when dry filled with wax. By this simple process, imitations have been prepared for the market sometimes passing for the genuine article.

The fish from which the scales are taken is so small that it is estimated four thousand are necessary to yield four ounces of the "essence."

Imitating precious stones by glass mosaics was an art understood by the ancient Greeks and Romans, which has been transmitted to their successors, and has become an important industry in modern Rome, Florence, Venice, and Sèvres. Venice, for many centuries, has enjoyed a monopoly for the production of aventurine, while the manufacture of imitation diamonds at the "Crystal Works of the Jura" is said to require the labor of a thousand or twelve hundred operators.

During the Augustan age and for two centuries later, the art of making paste jewels was carried on to a great extent, but, as a natural consequence, it went out of use on the decline of genuine work. It was not, however, exterminated, for at the Renaissance, when new vigor was imparted to all departments of art and learning, the occupation of making imitations became a profitable branch of industry, as may be inferred from the great number of spurious gems found in mediæval buildings and collections. Many of those used to embellish the churches of to-day are pastes, which have been substituted for the real gems to avoid any temptation for robbery. Those employed to ornament the sacred vessels of the Cologne cathedral are suspected of being imitations, as well as the onyx camei of the "Shrine of the Three Kings."

Stringent laws have been enacted at different times against counterfeiting gems; but in every instance they have been successfully evaded. And so long as they are prized as the most valuable of earthly possessions, frauds and imitations in their production and sale will continue to be practised.

Deceptions occur not only in the nature of the stone, but, also, in the manner of setting, by combining a genuine and a paste, or an inferior with a superior gem, as when a stone, cut as a "double," has the upper part garnet, for instance, and the lower glass, an artifice very difficult to detect. Garnets backed with crystal are, it is said, sometimes sold for rubies. To conceal a defect in color, the interior of the setting is painted or enamelled, which improves the tint, and gives it a beauty and intensity not inherent.

Tests. — Pastes may be recognized by certain trials, when properly applied. They will yield to the file, and are, therefore, deficient in hardness; they are liable to tarnish in impure air; they are not dichroic, as some real gems are. But the best test is afforded by specific gravity, which varies from that of real stones.

Some precious stones of a certain species bear a strong resemblance to others of a different species, as is the case with the pink topaz and the balas-ruby; hence, some convincing proof other than sight is necessary to distinguish them. A variety of experiments may be needed for this purpose. Take for an illustration any transparent, colorless gem, and test its hardness, to see whether it be a diamond. Can it be scratched by the sapphire, the next in the scale? If so, the stone is not a diamond. Here is a colorless gem, which looks very much like a diamond; how may we know it is not? Hold the crystal in a manner so that the rays of light shall be refracted, and you see a double image; therefore, it cannot be this precious stone which is single-refracting, and presents only one image; it may be a ruby, spinel, or garnet.

Let us select another specimen, so pellucid, and with so lovely a play of colors, that there seems to be no doubt as to its identity. Light is very deceptive; therefore, we will apply a pretty decisive test, that of specific gravity, and we find it to be 2.65, the same as rock-crystal, while that of the diamond is 3.5. But we have not done with it yet. What is the form of the crystal? It has six sides, and is called hexahedral. Now, the diamond never crystallizes in that form; its primitive crystals are octahedral. The conclusion is that the specimen in hand is rock-crystal.

We wish to ascertain whether a certain red stone is a ruby, and find its specific gravity less than 3.9; therefore, it must be some other gem. It is not clear that another precious stone, of a yellow color, may not be either a topaz or a jargoon, since both are found in this color, and both are electric by heat. We submit it to the ordeal of friction, and find it neither attracts nor repels these bits of paper; hence, it can be neither topaz nor jargoon, and we must make another assay.

The art of heightening or changing the color of precious stones is not a modern discovery, but was understood and practised by the ancients. Heat, as is well known, will produce this effect upon some gems; and to this agent is due the fine tints of the carnelian. It often effaces dark spots and impurities, and equalizes their color, a result which is secured either by wrapping the stone in a sponge for cremation, or by placing it in a crucible, and subjecting it to a high temperature.

The colors of precious stones are modified by a more complex process, which has been successfully accomplished with agates, chalcedony, and carnelian, at Oberstein and Idar, in Germany.

Combinations. — Precious stones of opposite or contrasting qualities should be placed near one another, in order to produce the most agreeable effect, as a step-cut beside a curved surface, a gem with adamantine lustre beside one with waxy lustre, and so of other contrasts. The diamond and the jargoon should not be in proximity; the former best harmonizes with the pearl or the cat's-eye, the latter with the turquoise.

Translucent gems, like the chrysoprase and the chalcedony, do not accord with chatoyant stones; while those reflecting the prismatic hues best associate with stones of less "fire," and those of one color with those of two or more. Rubies harmonize with moonstone; diamonds and pearls add to the beauty of pale-colored varieties, but contrast too strongly with those of deep tints. White sapphire, pearls, and jade, appear to best advantage in gold setting.

Antiquity. — Precious stones were used for various purposes by the earliest nations of antiquity, as we learn from history and tradition. They are frequently mentioned by the sacred writers as worn for personal ornaments, or employed for religious purposes, or as figures of rhetoric, to denote what is superexcellent in the realm of mind or matter. These allusions are not limited to one author or period, but they occur from Genesis to Revelations with more or less frequency, often constituting some of the most beautiful and striking metaphors to be found in literature.

Job speaks of the sapphire, onyx, ruby, topaz, crystal, coral, and pearls. He describes the process of mining, which was marvellously like that of the present day. Of one seeking for the precious substances found in the earth, it is said: "He putteth forth his hand upon the rocks; he overturneth the mountains by their roots; he cutteth out rivers among the rocks; and his eye seeth every precious thing."

Though the Israelites extensively used these costly treasures for both secular and sacred ornaments, they were equalled, if not surpassed, in these luxuries by their neighbors, the Phœnicians, judging from the practice of the Tyrian princes, who, according to the biblical account, displayed upon their persons no less than nine different species of gems.

The Phœnicians carried on a trade in these commodities, and are said to have introduced them into Egypt and Greece. If, as is supposed, they visited the Western Continent, may they not have imported them into Mexico and Peru, where precious stones were at a very early period used for similar purposes, if their traditions have any foundation in fact. There is evidence that both these nations understood the art of gem-engraving. Mexican seals and rings were set with precious stones engraved with the constellation of Pisces. The question arises, Where did they obtain their knowledge of Chaldæan astronomy? It has been said that these nations had no knowledge of the diamond; but, according to one of their traditions, a Mexican king, who was a poet, by the way, compared the sun to a "diamond, with a thousand facets," showing that he not only had a knowledge of this gem, but was also familiar with the modern art of cutting it. Their armor was jewelled in a manner similar to that of the knights of the Middle Ages, showing the Mexicans anticipated this feudal custom by many centuries.

Some antiquaries are of the opinion that the striking analogy between the jewels worn by these western nations and the Hebrews of Solomon's time points to the same origin for these races. With all of them, the emblems of sovereignty were the same. Their ecclesiastical and royal vestments were similar, both being covered with precious stones; their regalia were alike, embracing crown, bracelets, sceptre, sword, and other insignia. This parallel may be drawn between other nations, with similar results; therefore, it does not afford very decisive proofs of identity of race. The causes of striking resemblances between the customs of different nations in the use of ornaments undoubtedly have their origin in the universal love of the beautiful, and the desire to obtain what is most rare and costly for this object. There appears to be a pretty nearly uniform standard of taste as to the kind; they are generally for the head, ears, neck, arms, and fingers.

Precious stones were in general use in Homer's day; yet it

is an unaccountable fact that he seldom alludes to them, and only incidentally; while earlier and contemporary writers frequently mention them, as may be learned from the classic authors, from the Scriptures (already referred to), and from the traditions and literature of India, found in the great epics, the Ramayana, and the Mahabharata, parts of which were written, it is claimed, nearly four thousand years ago.

Very old Egyptian mummies have been found, decorated with crowns, necklaces, armlets, ear and finger rings, embellished with pearls and precious stones. Specimens of ancient gems, engraved with hieroglyphics, are to be seen in the Louvre, Paris. It has been thought that the interiors of the pyramids were once decorated with jewelled ornaments; it was probably on the tombs of eminent persons buried in these structures, a common practice in oriental countries.

Babylon has been represented as abounding in all manner of precious stones; and Damascus, in eastern metaphor, was a pearl encircled with emeralds, which proves the early use of gems for rhetorical figures, at least, and, by implication, for personal ornament. Some of the ancient jewels now contained in the British Museum, found among the ruins of Nineveh, bear date B.C. 700, and others are referred to a later period, including a bracelet inscribed with the name of Nimrod, B.C. 500.

CHAPTER II.

PRICES, TRADE, PAWNS, SUMPTUARY LAWS, ROBBERIES, AND SIZE.

Prices and Trade. — The commercial value of precious stones varies, like some other marketable commodities, according to the changes of fashion, and, like gold, they have risen and fallen in price by financial operations, by political changes, and by other adventitious considerations, but they have at all times constituted an important article of trade. It is stated that the price of diamonds fell fifty per cent when the interest on the debt of Brazil, due to England, was paid in that gem, and that in consequence of the political revolutions in Europe during 1848 and 1849, the diamond market received another check, which greatly affected their prices. No article of commerce was so sensitive to the instability of the market consequent upon the discovery of the New World, as precious stones; and their importation into Europe in immense quantities caused a panic among dealers, who endeavored to arrest this influx by representing them inferior to those brought from the East.

As an illustration of the fluctuating prices of gems, it is said that a cameo, with the portrait of Augustus and Livia, belonging to the Herz collection, which cost four thousand dollars, was sold, forty years later, for one hundred and fifty dollars.

It is probable that precious stones became an article of traffic at an early period of human history, even in the very infancy of nations, since they have always been highly valued as ornaments by all races. Emeralds were mined in Egypt fifteen hundred years before the present era; Palmyra, in the reign of Solomon, five hundred years later, was noted for its trade in gems, gold, and other valuable merchandise.

Before the discovery of the Western Continent, India, with some adjacent regions, was the great emporium for gems. Ceylon and Pegu yielded then, as they do at the present day, the largest supply of colored stones of the first class. They were a monopoly of the Kandy rulers, previous to the English control of the island, but now all restrictions are removed and no special grant to work the mines is necessary. The Malays, who are the principal dealers, cut and polish the gems for the Indian market, where they find eager purchasers in the native princes. The annual revenue from these mines is estimated at several thousand pounds. Brazil became a rival to India in the trade several centuries ago, before its separation from the control of Portugal, in 1822, and Lisbon, in consequence of her rich western possessions, led the world in the traffic in precious stones, which became a monopoly of the crown.

The relative value of first class gems has varied from time to time according to circumstances controlling the market. Statistics showing the comparative prices of the best known gems, selected from King, present the following: A perfect ruby exceeding one carat is worth more than a diamond of equal weight, and an emerald is worth four times as much, and, though the diamond has only doubled its value within a generation, the price of the sapphire has increased fourfold. The turquoise, like the diamond, increases its value in proportion to its size, while the chrysolite, amethyst, jacinth, and many other gems have no fixed price.*

* "The ruby to-day," says Mr. Kunz, "is worth five and even more than five times as much as the diamond, whereas the price of the emerald is rarely affected by the changes of fashion."

Pawns. — In the earlier history of nations, when wars were constant, and money, the sinews of war, scarce, sovereigns, who were as a rule, always poor and in sore need of funds, were obliged to resort to the expedient of pawning their jewels to obtain means for the prosecution of their ambitious schemes, or for maintaining their own power and dignity.

For ten centuries before the present one, it has been said that there were but few princes who had not pledged their crown jewels, and during almost the entire reigns of many of them they were constantly in pawn. This was the case of Edward III., who not only pledged his own jewels, but was willing that his queen should pawn hers, including her crown, to Flemish merchants, to raise money for him to prosecute his wars. This *chivalrous* prince allowed the Earl of Derby to go to a debtor's prison as a voluntary substitute for his sovereign. Richard II. placed the crown jewels in the hands of the Bishop of London and the Earl of Arundel, as security for money borrowed of London merchants.

From the Plantagenets to the close of the Stuarts, the crown jewels of England were frequently in pawn, in many instances on account of the personal extravagance of the reigning princes. Henry III., always embarrassed for want of funds, pawned his jewels, of which he had a large collection, to rebuild Westminster Abbey. Three hundred and twenty-four of his finger-rings, set with different gems, were at one time pledged to the King of France. This monarch should have been endowed with the hands of Briareus on which to display his ornaments.

Henry V. pawned the crown jewels, including the one known as the "Great Harry," and other regalia, during his wars in France, a part of which were subsequently redeemed, while Henry VI. and his heroic queen surrendered their most valuable treasures in the fruitless attempts to regain a lost throne. Henry VIII., James I., and James II., all used the same method to fill their coffers; the latter carried out of the kingdom many of the crown jewels as well as his personal ornaments, which were sold to various purchasers to obtain means of support during his exile.

These instances of royal poverty do not occur at the present day, and pawns of crown jewels are not in vogue, but vast sums are raised on diamonds and other precious stones by needy persons of both sexes among the high and the low classes of society. The Mont-de-Piété, established in Paris in 1777, an institution for the transaction of business of this kind, is said to have had in custody forty casks of gold watches at a time, which had been given as security for borrowed money. When the political troubles of the country suddenly reduced people of rank and wealth to beggary, the Mont-de-Piété was literally encumbered with valuable jewels which had been pawned by their unfortunate owners.

Sumptuary Laws.—Prohibitory laws against extravagance in the use of personal ornaments have been enacted, at one time or another, in nearly every civilized country, both ancient and modern; they were passed in Greece; they were promulgated in Rome, where the safety and even the existence of the Empire was imperilled by the luxurious habits of her citizens. Cæsar issued an edict forbidding the use of pearls for personal decoration except by individuals of a certain rank, and these only on days of public ceremony. His example was followed by some of his successors, but means were found to evade the laws, and the love of extravagance in the use of gems was stimulated, rather than checked, by these imperial decrees. The Emperor Leo, in the fifth century, issued the last prohibition against the excessive use of personal decoration by Roman citizens in the form of jewelry. Pearls, emeralds, and hyacinths were not allowed for baldricks and the trappings of horses, and men were forbidden to embellish the clasps of their tunics or mantles with precious stones of any kind.

After the fall of the Roman Empire, the trade in gems became obsolete until it was revived at the Renaissance. The early Christian writers condemned the extravagant use of jewels, and this sentiment has been cherished by some religious sects and individuals ever since. Anathemas and prohibitions against excessive luxury have by no means been confined to the church; secular writers have fulminated the most scathing satire against this human weakness, displayed by both sexes. Laws were at one time enacted in Florence prohibiting women from wearing jewels in public, and other instances are on record showing the opinions of lawmakers in regard to the influence of extravagance in dress upon the public interests of states.

The discovery and conquest of a new continent had a tendency to intensify the innate passion for ornament, throughout Europe.

In France this propensity received a temporary check in consequence of the sumptuary laws enacted during the reign of Charles IX., but they were ignored by the nobles, who carried their luxurious habits to a ruinous excess. The cost of a court dress was almost fabulous; nearly every article was loaded with pearls and precious stones. Though still more stringent regulations were imposed by Louis XIII. and Louis XIV., yet never before had extravagance been carried to such a height as during this period, and the use of precious stones exceeded all former examples in the history of that country, until it received a check at the Revolution.

The liberal use of costly jewels has not been confined to the laity, but has characterized the clergy as well. This tendency among the ecclesiastics during the Norman period in England was so conspicuous in the sacerdotal paraphernalia that it was deemed necessary to impose legal restraints upon it. Priestly vestments were at that time almost literally covered with costly gems. Thomas à Becket was a notorious example of this kind of clerical extravagance. The drinking cup of this ambitious priest, which has come down as a relic of mediæval times, is made of silver and ivory studded with pearls and precious stones. The high prelates generally did not fall below him in their fondness for personal ornaments, as is proved by the embroidered robes, covered with gold and gems, in which they were entombed. Chaucer and other poets have made the luxury of contemporary ecclesiastics a subject of keen satire and bold denunciation remarkable for those times.

Robberies. — The intrinsic value of precious stones, and the comparative facility with which they may be concealed or carried off, afford strong incentives for attempted robberies, which have in several notable instances been successful. Not even the sanctity and veneration attached to shrines, temples, and churches, have always preserved them from being despoiled of their immense wealth in these costly offerings.

In the reign of Edward I., Westminster Abbey, in which the royal jewels were deposited, was robbed of these treasures, though fortunately a large part of them were recovered. A bold but unsuccessful attempt was made by Blood, during the reign of Charles II., to carry off the regalia, which were kept in the Tower of London; and Queen Anne, consort of James I., was robbed of her personal jewels, valued at one hundred and eighty thousand dollars, of which no trace was ever discovered.

The church of St. Denis, in which were deposited ecclesiastical ornaments of immense value, was pillaged during the civil strife in France, the last of the sixteenth century, and again in the great Revolution of the eighteenth century. Mlle. Mars, the celebrated actress, was robbed in Paris, in 1827, of gems which, including their mountings, were estimated at ninety-six thousand francs, equal to nineteen thousand two hundred dollars, though the stones, which had been taken from their settings, were recovered. The Princess of Santa Croce, widow of an Italian prince, while residing in Paris, lost a number of valuable diamonds, which were stolen by professional thieves at the instigation of her lady companion and the Marquis of Loys. The parties, in this instance, were detected and punished, a retribution the robbers escaped who appropriated the jewels of the Princess of Orange, at Brussels, about the same time. In 1860, a robbery was committed at the Galleria della Gemma, Florence, when many valuable jewels were lost, including several engraved diamonds and rings of the Cinque-cento period.

The most notorious robbery of jewels that has occurred in modern times was the mysterious and astounding burglary committed at the Garde Meuble, Paris, in 1792, when a large collection of valuable gems and jewels, constituting the regalia of France, deposited in a large chamber of the Treasury, called the Garde Meuble, which was always strictly guarded, were carried off by some person or persons, who got access to the Treasury by climbing the colonnade of the Place Louis XV., and succeeded in escaping with the plunder. It had been the custom, before the Revolution, to exhibit these jewels occasionally to the populace, but after that event it was deemed prudent to close the deposit and affix seals to the cases holding them, a circumstance which rendered the affair of the robbery still more bewildering. A person confessed, several years after, that he had been one of the party concerned in the crime; but his communication was never made public, a circumstance which awakened the suspicion in the community that individuals of high rank were implicated in the theft. The most notable jewels stolen were three crowns, the sceptre and other emblems used in the ceremony of coronation, the golden shrine bequeathed by Cardinal Richelieu to Louis XIII., vases cut in agate, amethyst, and rock-crystal, the famous "golden rose," weighing one hundred and six marks, the Sancy and Regent diamonds, a rare blue diamond, the magnificent opal called the "Burning of Troy," a splendid brilliant afterwards recovered and worn by Napoleon I. at the battle of Waterloo, where it is supposed to have been lost, and a very large number of other gems of great value. Some of this plunder was restored by the robbers through fear of detection, by hiding the articles and then giving information where they were concealed. In this way, the Regent, the agate vase, and some others were recovered; but the larger part were never regained.

Subsequent robberies of the French jewels have been attempted; one in 1804, when the celebrated "Cup of Ptolemy," or "Vase of St. Denis," capable of holding more than a pint, and enriched with gold and gems, was taken from the Musée, at Paris, and, though recovered, it was first despoiled of its costly ornaments; and another in 1848, during the transfer of the crown jewels to the Treasury, when two pendeloques of diamonds, and a rare hat ornament of brilliants, were stolen.

Size.—The minuteness of precious stones compared with other articles of great value, instead of being a defect, as might seem at first thought, is really one of their merits, in certain respects, since they are more easily and secretly transferred, in case of emergency, from one place to another. An instance of this kind happened when the Prince Palatine, after the battle of Prague, in 1620, succeeded in carrying off his jewels, valued at a great price, with the proceeds of which he was enabled to defray his expenses during his exile in Holland; a parallel example is afforded in the history of James II. of England.

When a gem is spoken of as large or small, it is, of course, in reference to the size of others of the same species, and, in this sense, a precious stone may be called gigantic when of an unprecedented weight; there are a number of this kind on record. Among diamonds there are several of this class, as may be seen in the "Table on the Size of Celebrated Diamonds"; of sapphires a very large specimen, if not the largest, weighs one hundred and thirty-two and one-sixteenth carats, and is called the Ruspoli, the name of one of its owners, also the Wooden-spoon-seller, from the occupation of its discoverer, in Bengal. It was bought by a Parisian jeweller for thirty-four thousand dollars, and is now in the Museum of Mineralogy, Paris. The "Hope" pearl, forming a pendant in the imperial crown of Great Britain, weighs three ounces, or three hundred and sixty carats, and is considered the largest known, and a cat's-eye, called also the "Hope," measures one and a half inches in diameter.

The largest ruby known in Europe, presented to Catherine II., Empress of Russia, by Gustavus III., King of Sweden, when on a visit to her court, in 1777, is of the size of a hen's egg and of fine tint. The largest seen in India by Tavernier did not exceed fifty carats, and the largest in the French regalia is said to weigh less than nine carats. The Devonshire emerald from Bogota, South America, measures two inches in length and weighs nearly nine ounces; one owned by Duleep Singh is still larger, and a crystal from North Carolina has a length of eight and one-half inches. Austria claims an emerald of two thousand carats weight, an opal of seventeen ounces, and an onyx measuring nine inches in diameter. Nearly every museum comprises specimens of gem-minerals remarkable for their immense proportions, and some collections include engraved gems of gigantic size; in this list are found the "Cameo of the Vatican," on a stone measuring sixteen inches by twelve; the "Apotheosis of Augustus," or "Le Grand Cameo," in the French cabinet, on a stone of thirteen inches by eleven, and the cameo of Vienna, representing the "Coronation of Augustus," cut in a sardonyx of nine inches by eight. Not unfrequently various kinds of vessels of considerable magnitude are carved from a single gem, the quartz varieties affording some of the most remarkable for size.

CHAPTER III.

COLLECTIONS OF PRECIOUS STONES.

THE universal admiration for these treasures has led to the formation of valuable collections of gems by governments and by individuals, which have become the subjects of historical records; while among those who have been indefatigable collectors and connoisseurs are many celebrated names, both ancient and modern, — Alexander the Great, Mithridates, Julius Cæsar, Mæcenas, and Hadrian, of the former; and Frederick the Great, Napoleon, Goethe, the Dukes of Marlborough and Devonshire, among the latter.

Mithridates, King of Pontus, is said to have been the founder of the first royal cabinet of gems known to history, and M. Scaurus the first Roman collector. The immense quantities of precious stones brought to Rome by Pompey as trophies of his victories, awakened a public taste for these costly luxuries far exceeding that of any preceding era in the history of this nation; hence they were eagerly sought, not only for personal ornament, but also for enriching the cabinets of gem-collectors. These accumulations were called dactylothecæ, a word signifying finger-cases or boxes.

The most famous modern public collections of Europe are those of Paris, Florence, the Vatican, Naples, Berlin, Vienna, Dresden, Copenhagen, St. Petersburg, the Hague, and the British Museum and South Kensington in London. Among the smaller European collections, public or private, are the

Strozzi, Ludovisi, Antonelli, Castellani, Barbarini, Albani, Odescalchi, and the Collegio Romano, of Italy; those of the Dukes of Luynes and Blacas, Count de Portales, the Marquis de Drée, M. Fould, Baron Roger, of France; and the Devonshire, Northumberland, Marlborough, Townley, Knight, Rhoades, Maskelyne, and Townshend, of England. Several other collections have been more or less celebrated, as the Poniatowsky, Herz, Mertens-Schaffhausen, and Pulsky. Mr. Maskelyne's collection is said to excel in exquisite specimens of the glyptic art, as well as in the beauty of the stones themselves.

Collections in Great Britain. — No country in Europe, probably, is richer in antique gems than England. The British Museum contains specimens of the finest and rarest types of engraving on precious stones to be found anywhere, while some of the jewels comprised in this vast storehouse are of great age, dating from seven hundred years before the Christian era; but by far the greater number of these monuments of art, says King, are to be found in the cabinets of noble and wealthy amateurs.

The Museum includes the Blacas collection, which cost two hundred and forty thousand dollars; the Rhoades; the choicest specimens of the Castellani; and bequests from Messrs. Townley, Knight, and Cracherode, including in all about five hundred engraved gems.

Some of the Townley specimens, in the opinion of this writer, are unsurpassed by any from the most celebrated collections of Europe. They comprise many valuable Gnostic and Christian engravings, and some of the largest and most important antique pastes known. This institution affords some fine camei, including the head of Augustus, one of the largest of the kind, though the greater part of the engravings are intagli, accompanied by impressions in plaster, a great assistance in the study of this kind of work. Etruscan, Greek, Roman, mediæval, and modern art are all represented; the Etruscan antiques are in the form of scarabs, the Greek and Roman represent mythological subjects, while miscellaneous figures constitute the remainder. The Babylonian cylinders and Persian and Indian seals form an extensive and complete series of this class of engraved gems.

A few of the most celebrated engravings are Julius Cæsar, on sard, by Dioscorides; Livia, on amethyst; Perseus, on sard; Bacchus, on red jasper; a warrior and a dying Amazon, on amethyst; Cupid and Psyche, on sard; and a laughing fawn, on jacinth. A scarab on carbuncle which can hardly be distinguished from ruby is pronounced especially fine, but the Flora is considered by King, who is a skilled connoisseur in these matters, an imitation.

The South Kensington Museum comprises the Townshend and the Devonshire collections, both including valuable specimens of precious stones, and celebrated works of the carver's skill. Among the most remarkable of these objects of art are a cup made of oriental sardonyx of great beauty, inscribed with the date 1567, and the famous Cellini Ewer, once belonging to the French crown jewels. This pitcher, ten and one-half inches in height, is made of two convex pieces of sardonyx, with a foot of the same material, and a handle, stem, and spout of gold, embellished with enamel, rubies, and diamonds.

This collection at South Kensington contains also a second ewer cut in crystal, a Byzantine work of the ninth or tenth century; the largest known pearl, said to weigh three ounces, and set as a pendant jewel, besides other pearls of different colors and shapes; a large aquamarine, mounted in the hilt of a sword formerly owned by Murat; a cat's-eye obtained from the King of Kandy, and supposed to be the largest known, and numerous agates bearing natural representations of human features, and figures of different animals.

The Townshend collection, comprising gems of nearly every species and every variety of hues, many of which formerly belonged to the Hope cabinet, was bequeathed to this museum by Rev. C. H. Townshend, in 1869. It embraces one hundred and fifty-four specimens, nearly all mounted in gold, and forty-one engraved gems of both antique and modern workmanship. These comprise seventeen opals of different varieties, twelve sapphires of various colors, from violet to white, eight diamonds, including the rare black diamond cut as a brilliant, and others of honey yellow, pale green, gray, indigo, and cinnamon, rubies, emeralds, topazes, chrysoberyls, and specimens of a large number of other gem-minerals, affording a wide range of colors.

The Devonshire collection numbers five hundred and twenty-eight examples, including some of the finest antiques both in camei and intagli. It was made by the third Duke of Devonshire, during the first half of the last century, and has been augmented to its present size by his successors. In the list of these gems were numbered a fine amethyst engraved with the figure of Sapor I. and an inscription in the Pehlevi language, which now forms the centre ornament in the comb of the famous Devonshire parure; a Theseus, on sard; a Hercules, on green jasper; a muse tuning her lyre, on black jasper; Judgment of Paris, on onyx; a Marcus Aurelius, a head of Socrates, and one of Augustus.

The Marlborough collection is said to comprise the most extraordinary sardonyx known, on account of the color of its layers, which are purple, opaque white, and opaque black, affording a solitary instance of such a combination. The

Liverpool Museum contains a great variety of corals, including the Gorgonia or red coral used in jewelry, and Edinburgh has a collection of gems, many of them possessing an historical renown.

French Collections. — Many of the finest gems of antiquity have found their way to France through different channels; some by the acquisitions of her sovereigns, some by travellers encouraged by royal patronage, while others were the gifts of foreign princes or the spoils of war. Saint Louis and other Crusaders brought a large number of precious stones from the East; Tavernier, the most celebrated traveller of his time, added more from the same source, many centuries later; while Charles V. and his brother, the Duke de Berri, Francis I., Henry II., and Catherine de Medici, were all collectors and owners of vast stores of these costly treasures.

Charles IX. is said to be the first to arrange these accumulated gems in one collection, which was, however, dispersed during the public disturbances of his time, but re-established by Henry IV., who added others, and was planning to enhance its value still further, when his assassination prevented the accomplishment of his purpose; it was left to Louis XIV. to complete the work.

The Duke of Orleans, an enthusiastic collector, bequeathed his cabinet to the Royal Treasury, which was deposited in the Louvre and afterwards removed to the Bibliothèque Royale. The king purchased antiques from different countries of the globe, thus increasing the number gradually until it had assumed considerable proportions, and then removed the collection to his favorite palace, at Versailles. Other additions continued to be made, until the number of precious stones belonging to the crown was exceedingly large.

The intagli are distinguished for the beauty of the stones

and the variety of the subjects. A few of the most conspicuous are the signet of Michael Angelo; the "Apotheosis of Augustus"; the Agate of Sainte Chapelle, brought to France by Baldwin II. in 1244; the "Apotheosis of Germanicus," obtained at Constantinople and kept in the convent at Tours until presented to Louis XIV., in 1684; the Jupiter of the Cathedral of Chartres, and the Vase of Ptolemy or St. Denis.

The Louvre at present contains a large collection of gems, including the state or crown jewels, and numerous cups in rock-crystal, agate, onyx, and jasper.

In the Hotel de Cluny are seen many interesting relics of former ages: a set of chessmen in rock-crystal, formerly kept in the Garde Meuble, said to have been given to St. Louis by the "Old Man of the Mountains"; a bound volume embellished with precious stones, and the gold crowns, made in the seventh century, found near Toledo, in 1859.

The collection of Mlle. Mars was considered the richest owned by a private individual at that time. It comprised a very large number of brilliants and rose diamonds, pearls, topazes, emeralds, rubies, turquoises, corals, and camei. Many of the finest private collections are now found in Paris, including those of M. Turk, Baron Roger, the Duke de Luynes, and the Blacas. The Fouid cabinet was sold in 1860, in consequence of the death of the proprietor.

Italian Collections.—Lorenzo de Medici laid the foundation of the Florence collection in his cabinet of engraved gems, subsequently augmented by Cosmo, and other members of the family, who were all liberal patrons of art. It is known as the Florence "Cabinet of Gems," and embraces about three thousand specimens, including more than one thousand intagli and one hundred and eighty camei considered of superior excellence. The collection contains more than four hundred different objects cut in precious stones, including rock-crystal, lapis-lazuli, and others of a second class, all enriched with gems. Some of these articles are remarkable for their beauty and excellence of workmanship, comprising a casket representing twenty-four scenes from the life of Christ, made for Pope Clement VIII., by Belli Vicenza, and regarded the rarest work in the collection; a vase attributed to Cellini; a bas-relief in gold and jasper; a cup ornamented with pearls, representing a classical scene, by Bologna; a bas-relief in gold and gems in imitation of the Piazza della Signoria, one of the principal squares in Florence; besides numerous vases, cups, bowls, columns, and other objects, in different kinds of precious stones.

The intagli and camei represent a variety of subjects, and afford excellent studies in antique art; some of the most remarkable are the "Antoninus Pius," of extraordinary size; the "Judgment of Paris," a favorite subject for engraving; and "Hercules and Hebe." The ornamental tables in this collection are wonderfully beautiful and rich in gem-decoration, notably those of Persian lapis-lazuli. One of the number required the labor of fourteen years to make it, and is valued at one hundred and fifty thousand dollars. Others, made of jasper, and different costly materials, are all embellished with precious stones and pearls, representing mosaics, birds, foliage, flowers, vines, grapes, and shells.

The Vatican contains a great number and variety of precious stones, accumulated from time to time by chance acquisitions. Visconti made a catalogue of this extensive cabinet, which filled two folio volumes, but unfortunately it was lost before publication. The royal palace of Capodimonte, Naples, comprises a large collection of gems, while the palace at Caserta, once a favorite resort of the royal family, has been nearly despoiled of its works of art, though a few interesting specimens remain to attract the attention of the visitor. The cabinet at Naples, numbering between three and four hundred intagli, and two and three hundred camei, ranks second in Italy, the one at Florence being the first. It comprises the famous Farnese Vase, cut from one piece of sardonyx, and cost one thousand ducats.

Among the private collections are the Strozzi at Rome, which contains a Hercules, a Medusa, an Æsculapius, and a Germanicus, all works of merit as engraved gems, but many of the best productions of this cabinet have been transferred to the Blacas. The Ludovisi collection, belonging to Prince di Piombino, numbers many specimens of great value both antique and of the Cinque-cento period; its chef-d'œuvre is the Demosthenes of Dioscorides.

The collection of antiques exhibited by Sig. Alessandro Castellani, of Rome, at the Philadelphia Exposition, comprised various kinds of jewels both for personal ornaments and funeral rites, including engraved ring-stones used by the early Christians, all found among ancient remains, mostly in Italy. Among these relics were two amulets made of amber in the form of rams' heads, bearing date B. C. 700; a necklace of eleven amber cylinders set in gold, with six pendants in the shape of anchors; a necklace of the Roman imperial epoch, composed of sapphires, amethysts, and plasmas combined with blue glass cylinders and groups of leaves, Assyrian cylinders, Phœnician, Etruscan, and Greek scarabs, and amulets and rings of different kinds of engraved precious stones. Two jewels in this collection possess an historical interest, one of the number being a large ring made of gold, set with a garnet mounted on a pivot, and engraved with the portrait of Assander, King of Bosporus, which is considered one of the most remarkable of the kind known, and a second ring with red rock-crystal bearing an inscription and the arms of Pope Pius II. — 1458–1464. The engraved stones embrace a large variety, including the diamond, thus proving its use on the Eastern Continent before the discovery of America.

The Azara collection, owned by a Spaniard of that name, who lived during the last of the eighteenth century, comprises many camei and intagli of excellent workmanship.

The Berlin Collection. — This vast treasury of gems, one of the most valuable in Europe, was founded by Joachim I., an Elector of Brandenburg, and has been augmented from time to time by his successors, until it has assumed its present proportions, numbering five thousand specimens, including fourteen hundred set in rings and medallions. It includes the Stosch collection, comprising between three thousand and four thousand gems, the antique pastes of Batholdy and others, numbering from eight hundred to nine hundred, besides other acquisitions of a later date. The collection embraces examples of the Egyptian and oriental styles, dating from their most flourishing periods to 300 A. D., the oldest Greek and Etruscan gems, those of the later Greek and Roman periods, comprising intagli representing the gods, heroes, portraits, historical events, animals, and ancient monuments, illustrating the manners, weapons of war, utensils, etc., of that time. The camei include a large number, engraved with classical scenes to a large extent, and contain some of great size and value; as the "Apotheosis of Septimius Severus," on an onyx measuring eight and a half inches by seven, representing the Emperor and Juno drawn in a triumphal car by two eagles; this cameo cost twelve thousand thalers. The finest gems are mounted in gold, the remainder in silver, both with plaster casts beside each specimen.

The Stosch gems, named for their collector, who died in

1757, include antique and modern pastes, but the forgeries, says King, which have been added to the genuine works, have brought the whole group into discredit. The University of Berlin contains some fine gems.

The Herz Collection is a miscellaneous assemblage of precious stones gathered in imitation of the Stosch cabinet, and embraces every variety of subject without regard to the material or the excellence of workmanship. More than half of this "heterogeneous store" consists of pastes.

The Dresden Collection. — This "Historical Museum" enumerates among its treasures a very extensive miscellaneous accumulation of articles in precious stones, which are the progeny of a period extending from the close of the sixteenth to the beginning of the eighteenth centuries. It was begun in the reign of Duke George, about 1539, but the Elector Augustus, 1553–1586, was the first to deposit this accumulating wealth of art treasures in the apartments of the Saxon royal palace called the "Green Vaults." These rooms, eight in number, preserve one of the most unique collections of precious things found in Europe. The gems are numerous, and many of them are valuable for their historical reputation as well as intrinsic worth; but the lavish use of them to decorate ordinary and ignoble objects, is open to criticism; it is degrading to art and offensive to a cultivated taste.

These ornamented things comprise military weapons and defensive armor belonging to the Saxon kings, cups, vases, goblets, snuff-boxes, spoons, knives, cane-heads, drinking-horns, fruits, musicians, harlequins, dancers, peddlers, dwarfs, animals, and various other objects, all more or less decorated with precious stones and pearls.

From this bewildering mass, one can select specimens which afford interesting and curious studies illustrating the skill, ingenuity, and patience of the artist, and sometimes beauty in design. Here is seen a fire-place decorated with pearls and different species of precious stones; a monument constructed of corals, enamels, and gems; a grotto made of misshapen pearls; an oak cabinet covered with amber mosaics; portraits of the popes and emperors cut in gems; a mirror of rock-crystal; a ball twenty-two and one-half inches in circumference, of the same kind of stone, and a crystal beer-pot embellished with jewels and camei, valued at five thousand dollars. Court-dresses, royal trinkets, orders, decorations, chains, badges or favors, all loaded with gems, show the barbaric splendor of the Saxon court.

This museum contains a large onyx, measuring six and two-thirds inches by four and one-fourth, set in a gold crown, adorned with emeralds, diamonds, and pearls. One of the productions of Dinglinger, jeweller to Augustus the Strong, whose skill won for him the title of the "German Cellini," represents the Mogul Emperor of India, seated on his "Peacock Throne," surrounded by numerous courtiers and ambassadors paying homage to the great potentate, all executed in gold, enamel, and precious stones. This royal toy cost the artist eight years of labor, and the prince for whom it was made fifty-eight thousand four hundred and eighty-five thalers, or more than forty thousand dollars. There are more than four hundred different objects made of ivory, embellished with gems and enamel, and two hundred portraits engraved on gems. The diamonds are numerous, one ornament alone, for a lady's hair, comprises six hundred and sixty-two of these gems.

Vienna Collection. — It is said the development of the lapidary's art can be traced in this collection, from the fifteenth century to the present time. It embraces a large number of jewels of priceless value, including nine hundred and forty-nine intagli and two hundred and sixty-two camei; among them is the "Triumph of Germanicus," known as the "German Augustea." The most remarkable historical relics are the crown,* sceptre, imperial globe, sword, and coronation robe of Charlemagne, all profusely decorated with gems; the insignia of the Holy Roman Empire; the famous salt-cellar of Benvenuto Cellini, made for Francis I., King of France; a bouquet of gems, designed for Maria Theresa; a cameo, representing the Apotheosis of Augustus, comprising twenty figures; an onyx, nine inches in diameter, found in Jerusalem, by a Crusader, and sold to the Emperor Rudolph II. for twelve thousand ducats; an agate vase, measuring twenty-nine and one-half inches in diameter, the bridal gift presented to Mary of Burgundy; a goblet, in crystal, covered with precious stones, captured from Charles the Bold at the battle of Grandson; and a magnificent opal, weighing eighteen ounces, obtained from the mines of Hungary.

The Mertens-Schaffhausen collection was made by Madame Mertens-Schaffhausen, of Bonn, during the present century, from various sources, and is considered one of the best private cabinets of gems ever gathered. It numbered one thousand six hundred and twenty-six genuine stones, and more than two hundred pastes, and included the collection of Praun of Nuremberg, made in the sixteenth century, which consisted mostly of intagli, nearly all antiques. Half of these engravings are on sard, the remainder on chalcedony, topaz, amethyst, agate, onyx, and obsidian. The collection was sold in 1859, and new accessions were made, comprising some specimens from the Herz.

The Poniatowsky Collection. — Mr. King says, in all the cabinets of Europe taken together, there are not, certainly, one

* A sword and sceptre in the Museum of the Louvre, Paris, is said to have belonged to this celebrated hero.

hundred gems inscribed with the real names of the artists who engraved them. Many antiques — so called — have the names of ancient engravers added by a modern hand, with the view of enhancing their value. Probably the greatest forgeries of precious stones ever known were the Poniatowsky gems, all of which bear the name of some celebrated artist of antiquity — Pyrgoteles, Dioscorides, Solon, and others. These stones were engraved by the best modern artists of Rome, for Prince Poniatowsky, one of the members of the Polish family of that name, who died in Florence, in 1833.

The engravings are masterpieces, says this connoisseur, and, had the engravers affixed their own names, the gems would have increased in value with every succeeding age; whereas now they are regarded as comparatively worthless, and are sold merely for their gold mountings. At a sale of one hundred and fifty-four of these specimens, they brought only from twenty-five to thirty shillings apiece, though cut in the finest amethyst and sard, and set in splendid gold frames, of very elaborate design. Prince Poniatowsky inherited a valuable collection of genuine antiques from his uncle, Stanislaus, the last King of Poland, including some very celebrated intagli and camei, which renders it all the more surprising that he should have ordered one of counterfeits.

Russia. — Some of the finest and largest collections of gems in the world are probably found in the dominion of the Czar. They comprise more than ten thousand specimens, of which camei are the most numerous class, Egyptian, Etruscan, Greek, and modern works, many of them cut in rare materials, and inscriptions in the Coptic, Persian, and Turkish languages. The Museum of the Hermitage comprises many of the choicest specimens of some of the most celebrated cabinets formerly existing in Europe, including the Orleans, Strozzi, and others.

The School of Mines at St. Petersburg is said to afford a superb display of precious stones in their natural condition, obtained from the teeming mines of Siberia, comprising diamonds, emeralds, topazes, beryls, tourmalines, as well as many other species. Both the Winter and the Alexander palaces are depositories of valuable stores of these costly treasures, especially diamonds.

The United States. — There are no collections in this country, either public or private, that can compete with most of those of the Eastern Continent, in the number, variety, and size of the specimens, or in their value as historical gems, although nearly all the large cities, and many of the institutions of learning, have laid the foundation for valuable and extensive cabinets. There are several obvious reasons why we must not expect large and rare acquisitions of these treasures in this young republic, at least for some generations to come. Art collections are, generally, of slow growth. Those of the Old World are largely the accumulations of ages, and, in many instances, have been transmitted as a part of royal or titled possessions to legal successors, or they have been fostered and extended by State patronage, and owe their existence very largely to aristocratic governments, as is seen by the numerous collections of crown jewels; whereas, in a democratic government, they are principally the result of individual or associational contributions.

The best known public collections of gems in the United States are that of the National Museum, at Washington, which has been considered the most complete in this country; the one in the Metropolitan Museum of Art, at Central Park, New York; the collection at the Academy of Sciences, Philadelphia, and those belonging to some of the leading colleges and universities.

The Museum at Central Park is enriched by the collection of the "Curium Gems" brought from the Island of Cyprus by General Di Cesnola, and the King collection, comprising a series of engraved gems presented to this institution by its president, Mr. John Taylor Johnston.

The Cypriote Antiquities comprise a variety of objects interesting to the antiquary and the art student, including ear, finger, and seal rings bearing engraved stones, necklaces, bracelets, and armlets of curious workmanship, besides various other articles of ornamental use. One necklace is made of gold and rock-crystal, another of fine granulated work combined with gems. Two solid gold armlets weighing more than two pounds, were votive offerings presented by Eteandros, King of Paphos, whose reign dates from 672 B.C., with his name inscribed in Cypriote characters. The principal stones used for these engravings were carnelian, sard, rock-crystal, garnet, and onyx, and the subjects selected by the artists were generally those pertaining to Egyptian mythology.

The collection of Babylonian, Assyrian, and Phœnician cylinders engraved with various devices which have been reproduced on a flat surface afford an oppportunity for studying these ancient seals. Most of them are about one inch in length and represent in intaglio the beliefs, traditions, and customs of antiquity; the signet, or cylinder, of Nebuchadnezzar, King of Babylon, is among the number. Here are seen Egyptian scarabei with modern settings by Castellani, engravings supposed to date from the Twelfth Dynasty, between 2000 and 3000 B.C., intagli bearing representations of the Egyptian deities Ptah and Bast, and a relic taken from a mummy representing the cartouche of Amenophis I., who is supposed to have flourished 2500 B.C.

The valuable collection of antique engraved gems made by

Rev. C. W. King, numbering three hundred and thirty-one specimens, comprises examples of Greek, Roman, Gnostic, Assyrian, Phœnician, Etruscan, Persian, Indian, and Christian art. These engravings are executed upon a variety of precious stones, including the sard, jasper, lapis-lazuli, garnet, onyx, sardonyx, chalcedony, agate, nicolo, carbuncle, bloodstone, beryl, peridot, aquamarine, plasma, amethyst, rock-crystal, and emerald, but by far the larger part are on sard, mostly of an exceedingly beautiful and rare quality.

Many private cabinets, well known to amateurs, may justly claim the distinction of containing or having contained some fine specimens, but, unfortunately, a part of these have been broken up and the gems dispersed. It has been said that Dr. J. R. Cox, of the University of Pennsylvania, was one of the first gem-collectors in the United States, if not the first, laying the foundation for his cabinet nearly eighty years ago, and that the Leidy collection, which comprised more than two hundred specimens, gathered between 1860 and 1880, since sold and scattered, comprised some of the Cox gems. The collection of the late Rev. E. B. Eddy, of Providence, R. I., reckoned among the best, may possibly share the same fate.

It is not possible to decide upon the comparative merits of the various private collections in this country, nor would it be just to attempt it, since each has, undoubtedly, some marked peculiarity of its own, and must be judged according to its intrinsic worth. While a few of the best known collections are named, others of equal value may have been omitted. Of the former class may be mentioned that of Dr. Isaac Lea, Philadelphia; Mr. Lowell, Boston; Gen. G. P. Thurston,* Nashville;

* Through the courtesy and hospitality of General Thurston, an indefatigable collector of objects of *virtu*, and his amiable wife, the author was favored with the opportunity of examining his cabinet of gems, valuable for the number, size, and character, of the specimens.

Dr. A. C. Hamlin, Maine; Mrs. S. S. C. Lowe, Morristown, Penn.; Prof. C. U. Sheppard, New Haven; G. F. Kunz, New York; Mrs. M. J. Chase, Philadelphia; and Mr. Buffom, Boston.

Dealers in Gems. — The well known firm of Tiffany & Co., New York, is the largest, most comprehensive and complete in all its departments, of any establishment of the kind in the world, and has done much, says a contemporary writer, to give an impetus to the fine arts in America.

It has an international reputation for the beauty and originality of its designs, and the excellence of its workmanship, which are not surpassed by the best productions of London and Paris. As a mark of distinction, the honor of an appointment as special jewellers to a large number of foreign potentates and princes, has been given to this firm, the first of the kind ever conferred upon an American. About one thousand persons in all the different departments of labor, including the most skilful artists and artisans, are constantly employed in this immense establishment, and as a result the work of cutting, polishing, and engraving precious stones executed here, bears the marks of a scrupulous regard for the perfection of detail, and at the same time affords a pleasing general effect. Messrs. Tiffany & Co. have all the appliances for the delicate work of engraving portraits on the diamond, usually considered the highest achievement of the lapidary's skill.

CHAPTER IV.

CROWN JEWELS.

GEMS, it has been said, require the concomitants of royalty, grandeur, and beauty, to be appreciated in all their splendor and magnificence, and it is an historical fact that they have been universally employed as symbols of regal power in the decoration of crowns, sceptres, and other insignia of high rank. Royal crowns or diadems have a great antiquity, as supreme power invested in a single person became the prevailing form of government at an early period in the history of nations, though the sceptre, as a sign of royal power, has the priority of date. The first mention of a crown in the sacred writings is that of Saul's, which was brought to David after the defeat and death of the former.

The Crown. — This ornament was at first the emblem of the priestly office rather than that of the ruling power; but when the two prerogatives were united in the same person, it became more exclusively the representative of royalty. The crowns of the Egyptian, Hebrew, and Mexican rulers are thought to have borne a resemblance to the episcopal mitre of the present day.

The crown may have originated in the diadem, a fillet about two inches broad, worn across the forehead and tied behind, the two words, crown and diadem, subsequently becoming synonymous terms. At first, it was made of branches of flowers, more frequentl·· the laurel, the vine, wheat, etc., and afterwards of the precious metals, and, finally, it was garnished with all manner of precious stones until the crown became a "mine of wealth." This jewel varied in form among the ancients; that of the Mexican emperors consisted of a gold mitre elaborately adorned with feathers and precious stones, while among the Persians it consisted of a cap embellished with gems. The ceremony of coronation is not known in any part of Asia, says Tavernier. A cap adorned with the richest jewels is placed upon the head of the sovereign, but it bears no resemblance to a crown. The principal ceremony of investiture, both in Turkey and in Persia, is the girding on of the sabre; the same custom prevailed at the courts of the Mogul emperors and some of the native princes. Pearls seemed to be a royal favorite with orientals; a crown captured from the Tartars in the fifth century was profusely decorated with them, and a Persian crown, two centuries later, contained no less than one thousand large pearls.

The Emperor Heliogabalus adopted the pearl for his regal diadem, a style generally used in the Empire until the time of Constantine, when gold and different kinds of gems were used for the purpose. The emperors of the Middle Ages wore a diadem of silver when holding court at Aix-la-Chapelle, as king of Germany; one of iron when at Milan, as sovereign of Lombardy; and one of gold at Rome, as emperor of all their dominions.

Crowns, or, more properly, garlands or wreaths, were sometimes given for eminent services or talents, when they were designated by epithets signifying their character, as a triumphal crown bestowed upon warriors for signal victories, and those conferred upon successful athletes and poets, which usually consisted of ivy or oak leaves. From this ancient custom has descended the office of poet-laureate.

Scipio Africanus was honored with two hundred and thirty-four of these jewels to be carried in his triumphal procession, while those of Cæsar, on a similar occasion, numbered two thousand two hundred and eighty. Claudius, after his conquest in Britain, was presented with two crowns, weighing seven hundred and nine hundred pounds respectively, ponderous ornaments even for a sturdy Roman.

It is remarkable that with other numerous relics of antiquity, so few royal diadems have been preserved; indeed, scarcely any have come down to the present age except by historical records, until the discoveries near Toledo, in the middle of the present century, and, more recently, by the researches of Schliemann and other antiquaries, in Greece and Asia Minor.

The Spanish-Gothic crowns found in 1858 are monuments of Roman art after its decline. They comprise eight crowns and coronets of gold and precious stones, relics of the Gothic kings and nobles, and are now deposited in the Musée de Cluny, Paris. Some of these diadems are too small to be worn by an adult, and were probably intended for other purposes, it may have been for mortuary offerings at the tombs of the kings.

The largest crown belonged to King Receswinthus, who flourished about the middle of the seventh century, and consists of a circlet of gold embellished with large pearls, rubies, sapphires, opals, and emeralds, with the name of the king in gold letters suspended by small chains. Another diadem, similarly decorated has been referred to the queen, and others, less conspicuous for ornament, to the nobles; none of the gems were cut with facets. These valuable relics were discovered in a deserted cemetery at Fuente di Guerrazar, two leagues from Toledo.

Crown of Charlemagne. — This crown, used at his coronation at Rome as Emperor of the West, is set with large diamonds, emeralds, sapphires, and other gems. This symbol of imperial power, together with the gold throne, two gold shields, and other valuable treasures, was plundered from his tomb at Aix-la-Chapelle, in the middle of the twelfth century, by Frederick Barbarossa, and subsequently employed at the coronation of the German emperors; it is kept at Vienna as a relic of antiquity.

Crown of Hungary. — This venerable diadem, called the "Crown of St. Stephen," formerly used at the investiture of the Magyar princes with sovereign power, and worn by the first of this line of kings in 1072, was pledged by Elizabeth, Queen of Hungary, to Frederick IV., of Germany, in the fifteenth century. Accounts of its subsequent history vary; according to one statement, it has been kept in a place of concealment ever since the subversion of the Hungarian kingdom, while another places it in possession of the House of Austria.

An enumeration of the gems of this ancient crown gives the following summary: Fifty-three sapphires, fifty rubies, nearly three hundred and forty pearls, only one emerald, and no diamonds.

Iron Crown of Lombardy. — This famous diadem preserved in the Cathedral of Monza, Italy, dates from a very early period, possibly, before the sixth century, and figures also in modern history. It was used at the coronation of thirty-four Lombard kings, the Emperor Charles V., Napoleon I., in 1805, and of Ferdinand I., in 1838. This crown is not made of iron, as the name seems to imply, but consists of a broad hoop of gold embellished with different kinds of precious stones, and encloses a narrow circlet of iron, made, according to tradition, of a nail from the Cross, which was brought by the Empress Helena from Palestine.

The Sceptre.—This ensign of power, or its similitude, is very ancient, and its use to represent royal dignity and authority is thought to have been suggested either by the shepherd's staff, or one carried by persons of rank merely for ornament. Ancient kings sometimes bore a spear or javelin instead of a sceptre, as in the case of Saul, King of Israel, who, it is written, "abode in Gibeah, having a spear in his hand and all his servants" (courtiers and followers) "standing about him." It was this emblem of authority, the spear or javelin, with which the jealous monarch attempted to kill David, his rival. The sceptre of the Iliad consisted of a rod or staff made of wood surmounted by an ornamented ball or globe, as at the present day, and overlaid with gold, or adorned with gold studs and rings. This kind of sceptre is represented on the sculptured ruins of Persepolis.

REGALIA OF DIFFERENT NATIONS.

Austria.—The royal symbols of this empire are kept in the Burg at Vienna, and comprise several rich diadems and other insignia of regal authority, together with the private jewels of the imperial family. The crown jewels, which are numerous and exceedingly affluent in costly gems, include the crown and sceptre of Rudolph II., used by the German emperors on the occasion of their public entrance into their capital; the crowns of different Austrian sovereigns, resplendent with brilliant gems; decorations and other ornaments worn by the different orders, and the most complete collection of colored diamonds in existence, comprising the famous yellow Florentine, the Frankfort solitaire, which forms the centre of the Order of the Golden Fleece, and a pink diamond ornamenting the order of

Maria Theresa, or the Grand Cross. In this repository of precious things are crowns consisting of diamonds and rubies, others of diamonds, emeralds, and pearls, a bouquet of gem flowers, bracelets, and other ornaments, all displaying the riches of the mineral kingdom.

Dresden Museum. — The crown jewels include those belonging to the Saxon kings, together with what is sometimes called the Polish regalia, which, it is said, are not surpassed in brilliancy by any in Europe, and the crowns, sceptres, and globes used at the coronation of Augustus III. and his consort, Mary Josepha, in Cracow. Among the royal jewels * of this museum are crowns embellished with diamonds, emeralds, rubies, and sapphires, swords with hilts set with diamonds, aigrettes with pink and yellow brilliants, diamond buttons, and other gem ornaments for epaulets, hat-clasps and feathers worn by Saxon kings on state occasions, and the Orders of the Golden Fleece embellished with every species of gem. The jewels of the Saxon queens are no less conspicuous for their magnificence, including a necklace of diamonds, a shoulder-knot composed of fifty-one large and six hundred small brilliants, arranged about a centre stone, ear-pendants, hair-pins, and other feminine ornaments, all more or less garnished with precious stones.

Spain and Portugal. — The royal treasuries of these countries are especially wealthy in diamonds and emeralds, obtained principally from the New World, after the conquest. The Portuguese crown jewels comprise innumerable diamonds from the Brazilian mines, including the gigantic Braganza, if it is genuine, and several others of undoubted character, weighing between one hundred and three hundred carats. One of these gems, cut in the form of a pyramid, and set in the gold cane of John VI., has been valued at nearly one million francs, or two

* The crown jewels have been replaced by pastes.

hundred thousand dollars, while twenty diamond buttons worn in the doublet of Joseph I. have been estimated at a sum exceeding four hundred thousand dollars.

The Spanish treasury is no less replete with this precious stone, if the report can be credited that Queen Isabella II. displayed upon her person, at the public reception of an ambassador from Morocco, diamonds valued at two million dollars.

Russia. — This country is remarkably affluent in diamonds and other precious stones, which include many of extraordinary beauty and historical renown. The reason for this appears in the right of the crown to all the gems found in the productive mines of Siberia and the Urals, — a monopoly which has enriched the collection beyond the power of computation. Bayard Taylor refers to the great number of royal jewels seen at Moscow and St. Petersburg, and adds: "The soul of all the fiery roses of Persia lives in these rubies; the freshness of all velvet swards in these emeralds; the bloom of southern seas in these sapphires; the essence of a thousand harvest moons in these necklaces of pearls."

The geographical position of Russia has given her great facilities for the acquisition of oriental gems; while her conquest and absorption of smaller states have increased her stores of crown jewels to an almost unlimited extent. Uncounted wealth in precious stones, crowns, thrones, sceptres, globes, and other emblems of royal power, is deposited in the tower of the Kremlin, where the crowns of the czars and the regalia of the different peoples that constitute this complex nationality, comprising those of Siberia, Poland, Kazan, and other provinces, are kept. Some of the imperial diadems are exceedingly rich in diamonds, one alone comprising two thousand five hundred and thirty-six, all of superior excellence. This jewel is surmounted by an immense ruby, purchased at Pekin for one hundred and twenty thousand roubles. Two other royal crowns are embellished with diamonds, numbering from eight to nine hundred each. The crown of Vladimir, used at the coronation of the heir to the throne, brought from Constantinople in the year 1116, as a gift from the Emperor Alexis Comnenus, is ornamented, says Mr. Hamlin, with emeralds, rubies, and pearls; while his sceptre displays two hundred and sixty-eight diamonds, three hundred and sixty rubies, and fifteen emeralds. In this royal repository — the Kremlin — are safely guarded a number of other crowns, an orb, several thrones, one of them literally covered with gems, among which rubies and turquoises are conspicuous, and a second appearing like a glowing mass of diamonds, arms and armor, with all kinds of horse trappings, profusely decorated with precious stones.

Many of the imperial jewels are deposited in the Winter Palace, at the capital, comprising the crowns of the later sovereigns of the Empire, a beautiful diamond necklace, and other regalia. The crowns of the emperor and the empress are considered among the most magnificent diadems that ever adorned the brow of any potentate. The Hermitage, connected with this palace, constructed by Catherine II., is a vast treasury of rare and costly gems. This munificent princess not only poured out with a liberal hand these valuable gifts upon churches, palaces, and public institutions, but she gave to her favorite courtiers immense fortunes in this kind of wealth, comprising jewels of curious and unique style and workmanship.

Turkish Regalia. — The Ottoman Empire is supposed to be exceedingly opulent in crown jewels and precious stones, collected in the imperial treasury at Constantinople; but less is known of them than of those belonging to any other European court, owing to the more exclusive policy of the government. This vast repository comprises a large part of the valuable collections made under the Roman and Byzantine emperors, which fell into the hands of the Turks in the fifteenth century, and has since been greatly augmented by acquisitions from various nations conquered by or made tributary to the Turkish sultans. Some of these treasures have, however, found their way to France and other countries.

Permission was given, says Mr. Hamlin, to an English party, in 1840, and again to an American, in 1880, to visit the imperial treasury, and from these eye-witnesses we are surprised to learn that a country reputed to be so poor financially is so rich in costly jewels. A very few articles selected from this miscellaneous assemblage of royal emblems and garniture will give some idea of the variety and splendor of the whole collection.

Here are thrones blazing with diamonds, rubies, pearls, and gold, including the celebrated throne of Nadir Shah, costumes of the sultan bedizened with sparkling jewels, plumes with diamond fastenings, swords and daggers with hilts decorated with gems, shields elegantly wrought and jewelled, horse trappings, saddles and their coverings, embroidered with pearls and precious stones, knives, forks, spoons, and other articles of table service, clocks, inkstands, and snuff-boxes, all of them decorated in a similar manner. The imperial treasury holds a brilliant array of armor worn by the sultans; that of Murad II., the conqueror of Bagdad, is mentioned as being especially remarkable for its garniture of precious stones. A golden elephant standing on a pedestal covered with pearls, a table inlaid with topazes, the gift of the Empress of Russia, costumes trimmed with valuable furs and priceless gems, divans and cushions of gold tissue wrought with pearls, cradles of solid gold inlaid with precious stones, crystal vases encrusted with diamonds, rubies, and emeralds, and numerous other articles, equally rich in ornament, all afford proof of the former wealth, power, and magnificence of the Turkish sultans.

Persia. — The regalia of Persia, like those of Turkey, are reputed to be immensely rich in jewels and precious stones, which include several remarkable diamonds of great size and beauty. The royal treasury, not easily accessible to foreigners, was visited by Mr. Eastwick, who says the magnificent store of precious things valued at immense sums representing millions, were spread out on rich carpets to be examined. Various estimates have been made of the worth of these jewels, ranging from thirty or forty millions to two hundred millions, but all calculations made by strangers are, probably, mere guesses. A crown belonging to this collection is adorned with a ruby reputed to be of the size of a hen's egg, and a belt, weighing twenty pounds, is entirely encrusted with splendid rubies, emeralds, pearls, and diamonds. Necklaces of emeralds, aigrettes of diamonds, and other ornaments, with all kinds of armor, blazing with rare and costly gems, are to be found here.

The imposing ceremony of receiving foreign ambassadors at the Persian court, in Tavernier's day, was the occasion for all the nobles, high officers, and the royal horses, to appear in state, dressed in rich attire heavily ornamented with precious stones, while a Shah of later times, on a similar occasion, has been represented by an eye-witness as "refulgent with sparkling jewels." "He wore a tiara of the most brilliant diamonds, rubies, emeralds, and pearls; his robes were scintillating with glittering gems, strings of large pearls adorned his shoulders, his waist was encircled by a girdle of brilliants, while bracelets and armlets of remarkable beauty were displayed upon his arms and wrists." This glowing description will be credited after the Shah's European visit, when he astonished the English by the richness of his ornaments and the splendor of his equipage.

His coat worn on the occasion of his reception at the court of St. James, was all ablaze with brilliants of surprising beauty, five of these diamonds exceeding the Koh-i-noor in size. His sword, spurs, and decorations were covered with diamonds, rubies, and emeralds, while the caparisons of his horses exhibited the same splendid array of sparkling gems.

French Regalia. — It is not easy, nor, perhaps, possible, to estimate with any approach to accuracy, the relative value of the crown jewels of different countries. The palm has been awarded by some writers to France, and by others to Brazil; both are exceedingly opulent in gems of various species, some of which have become renowned for their historical fame. According to the inventory made in 1791, by order of the Assembly, the list embraced nearly ten thousand diamonds, and more than eleven hundred other gems, including pearls, rubies, sapphires, emeralds, topazes, amethysts, garnets, and others. The value of all the crown jewels made at the time, was estimated at nearly thirty million francs, or six million dollars, but a large part of this wealth was lost at the robbery of the Garde Meuble, soon after. Through the efforts of Napoleon I., on his accession to power, many of the original gems stolen from the Garde Meuble were recovered, and a large number of others were added to the collection, so that by the inventory of 1810, it presented the astonishing number of more than thirty-seven thousand specimens. The French crown is computed to contain five thousand two hundred and six brilliants, weighing in the aggregate nearly two thousand carats, and valued at a fabulous price.

Louis XVIII. and Charles X. added to the royal treasury great stores of precious stones, swelling their number to nearly sixty-five thousand, which were estimated in 1849 to be worth many million dollars.

It was noticed in the public prints, in 1881, that the subject of disposing of some of the crown jewels was discussed in the National Assembly, and in 1884 it was announced that they, or a part of them, were soon to be sold, and the proceeds were to be used in enriching the National Museum.

The most valuable jewels, after the regalia, are the decorations sent to the sovereigns by foreign potentates, comprising a watch presented to Louis XIV. by the Dey of Algiers, a brooch of diamonds, and a sword. It was decided these should not be sold, but others of no historical value were to be disposed of, including three parures, one of sapphires, one of turquoises, and the other of rubies, all of modern workmanship, made for the Duchess de Barri, but re-arranged expressly for the Empress Eugénie. It is said that the state jewels now comprise sixty fine diamonds, of more than twenty-five carats each.

Netherlands.—The Dutch imported large quantities of precious stones from the Western Continent after its discovery and colonization, and since the diamond-bearing regions of the East Indies have been accessible to them, their collection has become greatly augmented from time to time, either by purchase or by conquest, until the royal treasury at the Hague ranks among the largest on the continent of Europe, though it has the reputation of containing a large number of forgeries.

Brazil.—The imperial jewels of Brazil are exceedingly rich in diamonds, which have been valued at nearly twenty million dollars. The largest and finest gems obtained from the native mines have been appropriated by the crown; therefore the distinction of possessing the richest regalia has been sometimes awarded to this empire, but the specimens, however great their intrinsic value, have not the romantic associations which cluster about some of the oriental gems.

England.—The crown jewels of England were first deposited in the Tower of London in the reign of Henry III., and have generally been kept there, though a part of them were at one time retained in Westminster Abbey. During the Commonwealth, all the royal ornaments and that part of the regalia found in the Abbey, including the ancient crown of St. Edward, an orb, and sceptre, were sold by order of the Council of State, and scattered; but after the Restoration, copies of the lost jewels were made, still retaining the old names and styles, and added to the few of the original number which had been recovered. The present crown jewels are now secured in an iron cage in the Wakefield Tower, and strictly guarded. Both Professor Tennant and Mr. Harmon, author of the "Tower of London," locate the regalia in this tower, while Mr. Murray says they are deposited in the Bloody Tower; these towers are contiguous, which explains this discrepancy.

There are seven crowns in all, including St. Edward's, used at the coronation of all the sovereigns since his day. This diadem consists of gold embellished with diamonds, rubies, emeralds, sapphires, and pearls; the other crowns comprise Queen Victoria's, that of the Prince of Wales of pure gold unadorned, the crown of the queen consort set with diamonds, pearls, and other gems, the queen's diadem, made for the second wife of James II., adorned with large pearls and diamonds, the crown of Anne Boleyn, and another, of Charles II.

The crown of her Majesty Queen Victoria, made in 1838, by Messrs. Rundell and Bridge, London jewellers, is a crimson velvet cap bordered with ermine and embellished with gems taken from former crowns, together with those furnished by the queen's orders, comprising diamonds, rubies, sapphires, and pearls mounted in gold and silver. The diadem weighs 39 oz. 5 dwt., and has been estimated at more than half a million, which, considering the number and kinds of gems it contains, is undoubtedly below its real value. A summary of these is as follows:—5 rubies, 17 sapphires, 11 emeralds, 277 pearls, and 2783 diamonds, including brilliants, rose, and table: total 3093 gems. In the centre of a Maltese cross made of diamonds, adorning the front of the crown, is placed the celebrated ruby given to the Black Prince by Don Pedro of Castile, in 1367, and worn in the helmet of Henry V. at the battle of Agincourt, 1415. The cross surmounting the diadem, contains the sapphire set in the ring of Edward the Confessor, which tradition has endowed with marvellous powers.

The jewel-room holds various other emblems of royal power, as sceptres, bracelets, orbs, and other articles used at coronations. Saint Edward's staff of beaten gold, four feet seven inches in length, and surmounted by an orb and a cross, is carried before the sovereign on such occasions, and the regal sceptre, with a cross of gold ornamented with a large table diamond and other gems, is placed in the royal hands by the Archbishop of Canterbury, who officiates at this ceremony. This sceptre is two feet nine inches in length with the pommel ornamented by diamonds, emeralds, and rubies, and bears a rose, shamrock, and thistle, emblems of the different countries of Great Britain. The Rod of Equity, a sceptre surmounted with a dove, and embellished with a band of rose diamonds, is borne in the left hand of the sovereign at his investiture with regal power. Two sceptres, of smaller size, called the queen's sceptres, are adorned with precious stones.

The regalia include the following: the king's orb, set with pearls and large amethysts; the queen's orb, similar to the king's; the pointless Sword of Mercy, called the Curtana; two swords of justice, one for the state, borne before the sovereign, and the other for the church; bracelets, spurs, ampulla for holding the anointing oil and spoon for use in the ceremony of consecration, thought to be a relic of the ancient regalia; salt spoons; salt cellars, one of the number modelled after the White Tower; tankards, maces, baptismal font, and wine fountain, all more or less ornamented with precious stones.

The treasures of the Tower comprise four crowns made for Edward I. One of these royal diadems is set with rubies, emeralds, and pearls; another with Indian pearls only; a third with emeralds and rubies; and a fourth, used at his coronation, is of gold garnished with emeralds, sapphires, rubies, and large pearls. Here are seen a chess-board of Queen Elizabeth, inlaid with pearls and precious metals, a Roman shield of gold bordered with rubies, emeralds, and turquoises, a ewer of mother-of-pearl set off with gems, an amethyst engraved with Hebrew characters, cups enriched with gems, and jewels belonging to the Order of the Garter, dating from the time of Charles I., all bedecked with diamonds and rubies.

Scotland. — The crown jewels of this country, after the defeat of Mary Queen of Scots, were with difficulty saved from dispersion. "The Honors," as the crown, sceptre, and sword of state were popularly called, were concealed in Edinburgh Castle, and escaped being captured, while a large part of the Scotch jewels which fell into the hands of the English, and those coming into possession of native owners during the civil wars of this turbulent period, were recovered; among the latter class was that known as the "Great Harry." After the accession of James VI. to the English throne, the large diamond of this ornament was removed to adorn the "Mirror of Great Britain," described in the inventory of the crown jewels, in 1605, as containing: one table ruby; two large lozenge-shaped diamonds, one of which was called the letter H of Scotland; a faceted diamond, bought of Sancy; two pearls, and several small diamonds. A hawking glove belonging to this unfortunate princess was ornamented with fifty-two large pearls and numerous small ones, twelve rubies, and seven garnets.

The regalia, now in the Crown Room of the Castle, consist of the sceptre, the sword, the golden collar of the Garter, conferred upon James VI., by Queen Elizabeth, with a "George and Dragon" appended, supposed to exceed in splendor any other decoration of the kind, the crown of Robert Bruce, and other historical jewels. These regal symbols were all deposited in an oaken chest at the union of Scotland and England, in 1603, but were for a time lost sight of, and their disappearance was involved in mystery. No trace of the missing jewels could be found until they were accidentally discovered by Sir Walter Scott, in 1817.

Diadems of the Nobility. — Crowns at first were worn only by sovereign rulers, but about the tenth century they began to be assumed by the nobles as the insignia of rank; consequently, dukes, marquises, earls, viscounts, and barons, have their distinctive coronets, recognized by their different ornaments. Among the English nobility, the ducal crown is decorated with strawberry leaves, that of a marquis is distinguished by pearls between the leaves, while the coronet of an earl has the pearls set above the leaves. The diadem of a viscount is encircled with pearls, and that of a baron is adorned with only four pearls.

Some conception may be formed of the richness of the jewels in the possession of English nobles by the ornaments called the Devonshire parure, remarkable for the number and variety of the gems employed in its construction. The parure comprises seven different jewels — a coronet, necklace, bandeau, comb, stomacher, and bracelets, made of the most beautiful gems of the Devonshire collection, including the amethyst, sapphire, emerald, ruby, plasma, jacinth, sard, onyx, lapis-lazuli, garnet, sardonyx, and carbuncle; some of these gems are engraved; others are cut in the form of camei.

These jewels were designed for the Countess of Granville, wife of the English ambassador to Russia, to be worn at the coronation of the late emperor, and attracted the admiration, says Mr. King, of the imperial, princely, and noble families of the empire, accustomed though they were to the lavish display of the richest and most splendid gems the world affords.

CHAPTER V.

SECULAR USES OF PRECIOUS STONES.

THE love of personal adornment, as has been intimated, is universal; it is not limited to one period, nation, class, or sex, but is shared by all, though it exists in different degrees, and is manifested in different ways.

This innate passion for ornament seeks its gratification in the acquisition of whatever is considered the most beautiful and becoming for this object, according to the taste and cultivation of its possessor. The savage is contented to adorn his person with beads and feathers; while civilized man seeks the most valuable and attractive things in nature to augment his dignity and comeliness, therefore precious stones have been used for this purpose by all who could possess them.

Were the profusion of gems worn for ornament in the higher ranks of society a criterion, there would seem to be no lack of these coveted treasures; and, as a consequence, their commercial value ought to be very small. But it must be remembered they are almost the exclusive endowment of a few privileged classes, and have never been owned, to any great extent, by the masses. The social distribution of precious stones has always been limited; and, on account of their imperishable nature, they have very largely descended by inheritance, with the titles and estates of their proprietors, so that, with all the accumulations of the past, and the new accessions from recent mines, there is little probability that they will ever cease to be *precious* on account of their abundance and general use.

Ancient Uses. — The nations of antiquity were very lavish in personal ornaments, a practice which has been imitated by their successors with remarkable facility. Modern excavations made in the ruins of ancient buried cities have brought to light many interesting facts connected with the early use of precious stones in decorative art. That the Egyptians — one of the oldest nations of antiquity — made free use of gems for this purpose, is evident from the articles of jewelry found in their tombs, and from the pictured walls of their structures. The paintings at Thebes show they were used for money, since the subjected nations are there represented bringing their tribute to Egypt, in the form of precious stones, tied up in bags, made secure by seals. When Assur-banipal, the Assyrian conqueror, invaded Egypt, about the middle of the fifth century, B. C., according to his own account, he despoiled Thebes of a vast store of valuable treasure, including precious stones, which he carried off to enrich his own possessions at Nineveh.

Ancient "Egyptian jewelry," says M. Perrot, "is conspicuous for its richness of material and fine proportions, but so massive that it would seem the artists had borrowed their forms from their architecture." Various kinds of personal ornaments have been exhumed from the remains of ancient structures, similar to those worn at the present day, as necklaces, ear and finger rings, armlets, and bracelets, set with engraved stones, generally those of an inferior class, comprising the amethyst, carnelian, lapis-lazuli, turquoise, jasper, and others. The finest specimens now extant have come down from the great Theban dynasties long before the time of Moses. Jewels belonging to an Egyptian queen are seen in the British Museum; while the Louvre comprises many more of these interesting monuments, including ornaments found in the tomb of a son of Rameses II., called the "Great Oppressor," which assigns them to the age of the "Exodus," and a considerable number of rings engraved with the names of Thothmes, Amenophis, and other rulers of the eighteenth and nineteenth dynasties, covering a period from about 1600 to perhaps 1300 B. C. One of the rings of this collection is set with green jasper, bearing the figure of Thothmes II. engaged in a lion-hunt; another, with an engraved stone, and remarkable for its great size, — too large to be worn on the finger, — was undoubtedly used as a seal. The ring given to Joseph by Pharaoh may have been of this kind. A statue in the Louvre, supposed to be one of the oldest in existence, is represented with bracelets composed of twelve rings; while the Balouk Museum, in Egypt, contains a statue adorned with an elaborate necklace.

No engraved stones have come down from the earliest dynasties, according to the author just referred to; but the presumptive evidence is strong that the Egyptians understood and practised the art of engraving in these remote ages. There are no doubts, however, that gem-cutting was a branch of industry under the Empire, from about 1600 to 1150 B. C. Whether they employed the lapidary's wheel, or secured the result in some other way, it is pronounced by judges that they produced some very fine work in this line. In cutting, they made use both of intaglio and relief, though they appear to have had no correct knowledge of cameo.

The Egyptians, like their successors in art, understood and practised the method of making glass imitations, and many ornaments of very beautiful and elaborate workmanship of this kind have come to light. One of these jewels, in the form of a necklace, composed of four rows of glass beads with pendants of different emblems, probably for charms, is now in the Museum of the Louvre.

The finger-ring appears to have been a favorite ornament with this ancient people, who covered their fingers with them, even the thumb; their women indulged in ear-rings, armlets, bracelets, and anklets, more or less ornamented with precious stones, while necklaces were common to both sexes. The Egyptian regal head-dress, or diadem, called the pschent, worn during the period of the Empire, was formed by uniting the white crown of Upper Egypt to the red crown of Lower Egypt.

If we pass from Egypt to the great empires of Assyria and Chaldæa, we shall find in the remains of these extinct nations important relics showing the early taste of the human race for personal ornaments. In Sargon's palace at Kharsabad, which has afforded a rich mine of antiquities, there have been found cylinders, together with necklaces, armlets, bracelets, and other ornaments, made of carnelian, jasper, sardonyx, amethyst, and other gems of this class, cut in various forms, sometimes in double cones alternated with disks.* The neck ornaments of Assyrian princes were made of separate parts, each designed as an emblem, and the wrists of both gods and kings were adorned with massive bracelets; one of these, seen in the Louvre, must have been designed for the gigantic statue of some god or hero, since it measures five inches in diameter. "Assyrian jewelry," quoting Perrot, "is still heavier than Egyptian, large in design and brilliant in color; in its details, it has a power not unlike Ninevite sculpture, but it rarely has elegance, so important in that kind of artistic work."

The Phœnicians and Syrians were early accustomed to the use of precious stones, as we learn from the writings of Ezekiel,

* The oldest jewelry of these nations was of bronze.

who alludes to the great variety of these worn by the princes of Tyre. The Phœnicians, from their commercial habits, very early acquired a knowledge of the customs and arts of the nations with whom they traded, and, as they were apt pupils, they soon adopted the kinds of ornaments used by their contemporaries, the Egyptians, Chaldæans, and Assyrians, which pleased them. Not satisfied with being merely importers, they early acquired the art of cutting and engraving precious stones for their own use. The seal, employed by all the civilized peoples of antiquity, became indispensable to these merchants for legalizing their business transactions, and for this purpose they selected the cones and spheroids of Western Asia, and the scarabs of Egypt in preference to the cylinder, and were the first to affix seals to rings — a practice imitated by the Hellenic tribes. These seals are very numerous in their remains; one of them, cut on transparent agate, dates from the reign of Sargon, and another is referred to the seventh century, B. C. The principal subjects selected by the engraver were the figures of the gods, frequently with expanded wings to denote perpetual motion, the lion with some animal as his victim (a favorite motive of all the nations, "from the Mediterranean to the farthest limits of Persia"), and a king in a fierce contest with a lion, the king, of course, always coming off victorious.

The Phœnicians adopted the same mortuary custom of placing the jewels of the dead in their graves, practised by all the nations of antiquity, which accounts for the large numbers found in their tombs, comprising rings for the fingers and the ears, bracelets, necklaces, and other forms of jewelry, and the seal beside a male corpse, but no warlike implements, as was frequently done by other peoples, implying they were devoted to the peaceful pursuits of commerce, rather than the turbulent scenes of war. Many of their temples were richly ornamented with gold, silver, and precious stones, and as an instance of the luxurious habits of the kings of Cyprus, one of the Phœnician colonies, it is said that Cato sent to Rome the sum of £16,000 realized from the sale of these and other treasures taken from the royal palace. Jewels and engraved stones have been found in the region of northern Syria, occupied by the Hittites, a nation of biblical fame.

As imitations of precious stones were produced as soon as the art of making glass and tinting it was understood, spurious gems have been discovered in the remains of all the cities of antiquity, comprising those of Phœnicia, whose inhabitants were renowned for their skill in the manufacture of glass. They frequently combined real gems with their imitations in the same jewel, as may be seen in bracelets, bangles, and necklaces of Phœnician workmanship, and sometimes personal ornaments were made of enamelled porcelain, and even of iron, which was regarded by the ancients as a precious metal.

Greece and Rome.—The use of precious stones for decoration was exported into Greece from Asia, thence into Rome; but the Greeks made a vast improvement in the works of their masters, by introducing human forms in jewelry and gem-engraving, which they did with all the refined taste and remarkable skill and ingenuity characteristic of this gifted race. Alexander, after his conquest, adopted the oriental fashion of wearing ornaments and robes displaying a profusion of gems, while the Romans not only imitated the Greeks, but vastly exceeded them, in the use of these expensive luxuries, which they employed liberally to adorn their own persons as well as to embellish their equipages and household utensils, and for other common purposes. In their wars with Mithridates, King of Pontus, who, perhaps, surpassed all other princes of ancient history for his acquisitions in gems, they obtained vast stores of precious stones and costly jewels. Lucullus, after his victories in Armenia, returned to Rome laden with rich spoils which graced his triumphs, including, among other trophies, a gold statue of the conquered king, six feet in height, bearing a shield covered with jewels, and so large a number of splendid vases, made of gems, as to require a car, drawn by camels, to transport them. A few years later, Pompey, after his successful campaign against the King of Pontus, was accorded a magnificent triumphal entrance into Rome, when he wore a mantle embroidered with gold and precious stones, and was attended by a procession bearing magnificent trophies of his military successes, including thirty crowns made of pearls, the throne, sceptre, and chariot of Mithridates, the latter formerly used by Darius, a chess-board with all its pieces of gold set with gems, and a gold vine with leaves and fruit, made of these rare productions, valued at nearly half a million dollars.* The diadem and scabbard of this Asiatic despot have been represented as a mass of splendid gems. Pompey, by his foreign wars, enriched the treasury at Rome in gold, silver, and jewels, to the amount of twenty thousand talents, probably about three and one-half million dollars. The triumphal entrance of the Emperor Aurelian into the capital, after his capture of Palmyra, was the occasion of another brilliant display, when Zenobia, the unfortunate queen, wearing a diadem and royal robes resplendent with costly gems, was compelled to pay reluctant homage to her unrelenting foe.

Julius Cæsar, who was an enthusiastic collector of precious stones, left immense treasures in this kind of wealth, which some of his successors applied to unsuitable uses; as when Caligula employed them to ornament the trappings of his favor-

* Mr. Jones thinks the golden vine a fiction.

ite horse and the sterns of his ships, and when Nero used them to add lustre to the panels of mother-of-pearl in his "Golden Palace." Herodotus describes a golden vine, bearing grapes of precious stones, placed over the couch of Pytheas, the Lydian king, and Quintus Curtius, in a glowing picture of oriental luxury, says persons of rank and wealth are conspicuous for their ornaments of precious stones, and when the king appears in public, he is borne on a litter adorned with magnificent pearls, while his palace is enriched with columns of gold, garnished with golden vines bearing fruit of brilliant gems, and supporting silver birds. The troops of Darius, called the "Immortals," when equipped for battle, wore jewels of gold and precious stones, while the king's chariot and armor were literally covered with them.

The excessive use of gems for personal ornament was severely criticised by contemporary writers. Pliny says Lollia Paulina, wife of the Emperor Caligula, sometimes appeared adorned with jewels worth immense fortunes. On one occasion, an ordinary wedding dinner, she was nearly covered with them, —head, ears, neck, and fingers glistened with costly jewels, valued at two million dollars. Writing of the extravagance of Roman ladies, Tertullian says: "The slight lobes of their ears outweigh a whole year's income, and their left hands squander a money-bag on every joint." A ring worn by Faustina, consort of Heliogabalus, was estimated at two hundred thousand dollars, and a pair of garters, with cameo fastenings, at nearly the same price; while the ear-rings of Calpurnia, the wife of Julius Cæsar, were valued at the sum of one million two hundred thousand dollars. The other sex displayed their extravagant tastes in a different way, as when the Emperor Heliogabalus entertained his guests with dishes served up with gold and precious stones, intended, it is presumed, as gifts to the company.

The excessive use of ornaments by the women of Rome aroused the remonstrances and condemnation of the sterner sex, who, of course, had the expenses to defray, and one burdened husband declared he would cut off the ears of any daughter who might be born to him, in order to save his future son-in-law the ruinous cost of ear-rings. Seneca said that "they" — the women — "are not satisfied with one pearl in each ear, but they must have three." These aggrieved husbands and fathers had good reasons for remonstrance, if it was true that immense fortunes were spent on ear-rings and other jewels, as appears to have been the case with some of the Roman ladies. It is said the fair owners were liable to severe injuries from the weight of these costly pendants, and that a special vocation existed at Rome, having for its object the healing of their ears. This luxury of wearing jewels in the ears was prohibited to men by an edict of the Emperor Alexander Severus, 222–235 A. D., which proves they were worn by the male sex before his reign. Precious stones were used for architectural decoration by the nations of antiquity, as they have been by those of the Middle Ages; the Ptolemies afford an illustration of this custom. The poet Lucan describes the luxury and splendor of the palace of Cleopatra, which would seem like a fabric of the imagination were not his narratives authenticated by contemporary history. Pavements of onyx, thresholds of doors made of tortoise-shell set with emeralds, furniture inlaid with jasper, and couches studded with various kinds of precious stones, met the bewildered gaze of the Roman soldiers who invaded Egypt under Augustus.

The Goths. — The excessive use of gems indulged in by the Romans was early adopted by the swarming tribes that conquered the Empire and occupied her territories. The victors carried off an innumerable amount of beautiful vessels cut from precious stones and profusely embellished with them, besides vast stores of costly jewels. It is on record that one dish, weighing fifty pounds, ornamented with gems, was presented to one of the Gothic kings by a successful general, and fifty basins filled with these valuable spoils were given to one of the princesses, a statement corroborated by the immense treasures in jewels found in the palace of the Visigoth kings at the pillage of Narbonne by the Franks, in the sixteenth century, and in the Gothic treasures of Toledo, presented in the form of crosses, chalices, pateras, caskets, and other articles, all elaborately garnished with precious stones, which were seized by the Moslem conquerors and sent to Damascus during their wars in the Peninsula. A remarkable discovery of buried treasure has also been made in Roumania, the ancient Dacia, a country occupied by the Goths, comprising gold and jewels of Gothic and Byzantine workmanship, probably the spoils captured by these warlike tribes, in some of their predatory excursions. The early Gauls cared less for personal adornment than the Franks, who, in the time of Dagobert, displayed their love of ostentation by the free use of gems to decorate their attire and weapons of war, but after these tribes embraced Christianity, their use of these ornaments was directed into another channel — that of embellishing their churches and other religious buildings, which, in the mediæval period, became depositories of wealth in gems and costly jewels.

After the reign of Charlemagne, there was a decline in the use of precious stones, so that in the twelfth century their possession was confined almost entirely to princes and ecclesiastics.

The Anglo-Saxons participated in the general passion for ornaments, embroidered robes, and crowns brilliant with precious stones, while the caparisons of their horses, with their richness of decoration, offered a tempting prize to their Roman conquerors. These relics are occasionally exhumed; Mr. Jones relates an instance of the kind which occurred in 1771, when a crown of complicated design was discovered on Kingston Down, and another was found in London, in 1840, composed of gold filagree and pearls.

Modern Uses. — Precious stones have been more generally employed for personal ornament in modern times, especially during the past and present centuries, than they were among the ancients, for the reason that there are a greater number of people of wealth in the private walks of life than at any previous time in the history of the race. The oriental nations have always been attracted by gorgeous apparel, and it is among them we shall find the rarest collections of gems and the most brilliant exhibition of their use for decoration. Prominent among eastern countries for richness of ornaments may be classed the dominions of the sultan of

Turkey. — At the time of the Turkish conquest, Constantinople was a vast store-house for the rarest and most valuable spoils of the world. Whatever was most beautiful in art, whatever was most precious in nature, had drifted thither, and this fact explains why so many remarkable and costly gems are owned by the Imperial Government or by the Turkish nobility. It is said that the extravagant use of these ornaments by the latter, during some periods in the history of this nation, was incredible. One of the pashas, at his death, alluded to by Mr. Hamlin, left thirty-two cuirasses studded with rubies, fifteen strings of large pearls, besides sixty bushels of small pearls, and numerous articles covered with diamonds, while an officer of state adorned his garden at his country-seat with parterres of flowers composed of gems in imitation of natural vegetation. It is hardly necessary to refer to the richness and abundance of the sultan's personal ornaments, as it must be taken for granted that a luxurious despot would not be surpassed in magnificence by his vassals.

On state occasions, as is the oriental custom, he is surrounded by all the symbols of rank and power. When the ambassador from the Dutch Republic, at the beginning of the seventeenth century, was granted an audience, he found the emperor seated upon a throne blazing with diamonds, rubies, pearls, and other costly gems, under a sumptuous canopy similarly adorned.

Tavernier, who visited the sultan's court, describes the splendor and magnificence of the Grand Signor's royal palace and equipage with the minuteness of detail characteristic of this writer: —

Eight different coverings were used for the drapery of the imperial throne, — one of black velvet embroidered with large pearls, another of white velvet ornamented with rubies and emeralds, a third of purple velvet decorated with turquoise and pearls, while the remaining five were embellished with gold and other rich materials. Different draperies were employed on different occasions; for example, on the reception of ambassadors, the richest were used for the representatives of the most distinguished and powerful nations, while for those of less consequence, in the estimation of the sultan, a more simple cloth of state was displayed. On these public occasions, the emperor's horses, consisting of a numerous retinue, were paraded in caparisons bedecked with diamonds, emeralds, rubies, and other precious stones. The magnificence of the court was carried into all its amusements, which afforded occasions for the display of the wealth of the Ottoman Empire in these treasures. Every sporting hawk wore a hood embroidered with pearls, and a jewel ornamented with gems adorned his neck. The imperial exchequer appears to have been an inexhaustible mine of diamonds, if it was true, as stated, that the Emperor Murad V., who acceded to power on the dethronement of Abd-ul-Aziz, in 1876, paid his banker four million dollars worth of these gems, which were sent to Paris for sale.

Mogul Rulers. — The luxurious habits of the Turkish sultans were imitated by *all* the Mohammedan princes, and perhaps surpassed, by the Mogul emperors of India, whose affluence in precious stones was inconceivable. The accounts given by travellers who visited their court almost surpass belief. Tavernier, whose opportunities for judging were not excelled, says there were in all seven imperial thrones, all of them literally covered with diamonds, rubies, emeralds, and pearls, but for richness and novelty of design, the famous "Peacock Throne," of Shah Jehan, exceeded all others. The outspread tail comprised sapphires and other gems to represent the natural plumage of the bird; the body was of enamelled gold studded with precious stones, while from the breast was suspended a large ruby with a pear-shaped pearl of fifty carats weight, for a pendant. The peacock was placed over the throne, and on each side were arranged bouquets of flowers made of gold and gems. The throne itself, six feet in length and four in width, was of solid gold covered with precious stones, including more than one hundred rubies, weighing from one hundred to two hundred carats each, and one hundred and sixty emeralds, from thirty to sixty carats in weight.

A transparent jewel with a diamond pendant, of eighty or ninety carats, encircled by rubies and emeralds, was suspended before the emperor when he occupied this imperial seat; twelve pillars, which upheld the canopy, were set with rows of pearls, and on each side was displayed a crimson velvet umbrella with pearl fringe and handles encrusted with diamonds. Behind this magnificent royal pavilion was erected a smaller throne covered with diamonds and pearls.

The treasury of Shah Jehan, the most celebrated of the Mogul rulers, was hardly equalled, for priceless jewels, by that of any potentate, of his own or any other period. His robes, sceptre, sword, shield, and dagger were encrusted with gems; three necklaces of huge pearls, bracelets of diamonds and other costly gems, crown, and turban bearing heron feathers confined with an immense ruby, diamond, and emerald, formed a part of his royal vestments. His tent, when in camp, was covered with rich scarlet cloth lined with purple satin embroidered with pearls, diamonds, rubies, amethysts, and other precious stones. One of the curiosities of his collection was a globe covered with gems to represent the different objects delineated on its surface, as the sea was designated by emeralds, and the various countries by other kinds of colored gems. The golden vine overhanging the portico of the Moslem palace at Agra was covered with grapes of emeralds and rubies, to represent their different stages of maturing.

Aurungzeeb, the son and successor of Shah Jehan, accumulated an astonishing mass of precious stones and rare curiosities, comprising, with many others, a table and cabinet, made of different colored gems arranged to represent birds and flowers in their natural tints. Mahmoud, the famous Moslem conqueror left, at his death, four hundred pounds avoirdupois in precious stones plundered from his vanquished foes.

The celebrated Tartar conqueror, Timour or Tamerlane, of the fourteenth century, acquired vast stores of precious stones during his campaigns in India, Persia, and Arabia, which he was wont to display with royal magnificence, during his sumptuous fêtes in Samarkand, his capital.

India. — The natives of India have always exhibited an

extreme fondness for precious stones, comprising some of the most splendid diamonds, pearls, rubies, sapphires, and emeralds anywhere to be found. The treasury of Tippoo Sahib, after his defeat by the English, in 1786, was found to contain jewels estimated at the enormous sum of four hundred million dollars. The Prince of Baroda, a most zealous collector, possessed uncounted treasures in gems, including many of European celebrity.* It has been said that the jewels of an Indian belle aggregate more than thirty pounds in weight. The dress of oriental ladies was always expensive, and has not changed materially in that matter since the times of the Hebrew prophets, by whom they were censured for their extravagance in ear-rings, finger-rings, chains, bracelets, nose-jewels, head-bands, tablets or pendants, and anklets, though nearly all the same kind of ornaments formed a part of male attire in ancient times, as represented by the sculptures of Persepolis and other ruined cities.

The aboriginal inhabitants of the Western Continent cherished the same fondness for personal ornament that has always characterized the nations of Asia, if we are to judge by the immense quantity of precious stones brought from Mexico and Peru by the Spaniards; nor have Europeans fallen much behind either the eastern or western races in the use of these, the richest of nature's material gifts, and the most beautiful of all her creations.

The House of Burgundy, from Philip the Bold to Charles the Bold, the last prince of the line, exceeded all other European sovereigns of that period in the richness and abundance of their gem-collections, especially of their diamonds. At

* The profusion and beauty of Indian gems exhibited at the London Exposition of 1851, says Professor Tennant, were almost bewildering. The girdle of a rajah was an object of general interest for its remarkable display of magnificent emeralds mounted without faceting or polishing.

the marriage of one of the scions of the family, the bridegroom presented gifts to each of the guests, consisting of diamonds, valued, in the aggregate, at nearly eighty thousand francs.

Philip the Bold, on great public occasions, appeared in jewels worth a fortune. His costume at the meeting with the Duke of Lancaster, of England, at Amiens, in 1391, as described by the chroniclers of the times, comprised a surcoat which was embroidered on the left sleeve with a branch bearing twenty blossoms made of pearls, rubies, and sapphires, and buds consisting of pearls. Other articles of his elaborate wardrobe were decorated with the same profusion of rich ornament, including a suit of crimson velvet, covered with rubies and sapphires, which was to be worn with chain and bracelets set with rubies. When regent of France, this luxurious prince indulged in the same reckless extravagance, and, as a consequence, died a bankrupt. Philip the Good surpassed his predecessor in the splendor of his ornaments and the magnificence of his retinue. His collection of gems was so large and varied that he is said to have worn at public receptions jewels valued at one million francs, and to have changed them each day of the week, so that, if diamonds were the choice of one particular day, rubies were selected for the next, sapphires for the third, and so on to the end of the week, according to the fancy of the owner. As a matter of course, his example was followed by the numerous suite that always attended him, making his court one of unparalleled splendor and luxury.

Charles the Bold, the last Duke of Burgundy, exceeded all the princes of this house in his acquisitions of precious stones. His ducal crown of pearls and diamonds was "worth a whole duchy." A mantle worn by this prince, covered with gold and diamonds, cost two hundred thousand ducats; while the tent used in his campaigns was remarkable for the richness of its ornaments. The duke's escutcheon displayed on the outside of the tent was embellished with pearls and precious stones, and the interior was hung with crimson velvet, embroidered with golden foliage and pearls. The throne which formed a part of his military equipage was of massive gold, and the hilts of his sword and dagger were covered with rubies, sapphires, and emeralds; while his cap of maintenance and plume-case were enriched with diamonds, pearls, rubies, and sapphires. His Order of the Golden Fleece and seal were garnished with precious stones of great beauty and rarity. Some of the jewels of this duke have a place among the gems of literature, including the "Lamp of Flanders," the "Three Brothers of Antwerp," consisting of three rubies, and, according to some writers, the Sancy diamond. A large part of these jewels were captured by the Swiss, at the battle of Grandson, and scattered throughout Europe; some of them belong to the French regalia. Sir Walter Scott, in his novel of "Anne of Geierstein," gives a graphic account of the wealth of this prince in costly jewels, and of their loss during his unfortunate war with Switzerland.

The Duke of Orleans, who flourished during the latter part of the fourteenth century, was distinguished for his munificent gifts in precious stones to his favorite retainers, as well as for his own luxurious habits of living. He displayed his love of ornament not only in his personal attire, which was richly decorated with brilliant gems, but also in all his domestic arrangements; his gold and silver plate, and other articles of table service, being inlaid with them.

The Italians and Spaniards of the fourteenth and fifteenth centuries emulated the other European states in the amount and value of their gem collections; while their nobles displayed on their court-dresses whole mines of wealth. The mantilla of

Catherine of Aragon, at her marriage with Henry VIII., was ornamented with a border twelve inches deep, covered with gold, pearls, and other costly gems. The ducal crown of one of the Visconti, a noble Italian family, was set with precious stones estimated at a great price for those times.

France. — The partiality of the French for the extravagant use of personal ornaments had declined after the time of Dagobert and his immediate successors; but Louis XII., on his accession to power, made an attempt to revive the national taste, by inviting jewellers to France, from Milan and Genoa, and succeeded so well that, during the reign of his successor, Francis I., precious stones became objects of eager pursuit by both sexes. Consequently, the art of setting gems was carried to a high degree of finish, a distinction the French have ever since maintained. Many of the designs were made by some of the most celebrated painters of the times; several were executed by Cellini, the most famous artist of his age. Colored stones and pearls held the supremacy until superseded by diamonds, in the last half of the eighteenth century. The extravagant use of precious stones in France at this period exceeded all bounds, and rivalled, if it did not surpass, that of the ancient Romans at the height of the fever. The dress of Marie de Medici, at the baptism of one of the royal children, was trimmed with thirty-two thousand pearls and three thousand diamonds, valued at sixty thousand crowns; her robe was so heavily weighted with these ornaments that the queen was not able to bear the burden without assistance. Louis XIII. and his courtiers exceeded all their predecessors in extravagant luxury, insomuch that an attempt was made to check it; but fashion will not tolerate any interdict, and various methods were employed to evade the law. What decrees could not do, was finally accomplished, for a time, at least, by a political convulsion; the Revolution arrested this reckless prodigality, to be revived, however, at a later date, with still greater vehemence. The era of classical styles followed, and costumes were elaborately trimmed with precious stones, and all kinds of jewelry were modelled after the antique. The number of fingers were too few to display all the rings considered necessary to set off personal charms; therefore, the toes were brought into requisition. But here was another difficulty: how could these glittering ornaments be displayed with the feet covered? Fashion is very inventive; so the feet must be dressed in oriental sandals, in order to exhibit their fine jewels. The noble families exiled by the new government had taken their hereditary jewels with them, which left only the less valuable gems; but these served the purposes of camei, hence camei were greatly admired until the banished diamonds and pearls re-appeared in the circles of rank and wealth.

The English princes and aristocracy adopted the habits of their continental neighbors in the display of personal ornaments, especially during the Norman period, when kings, nobles, and prelates were conspicuous for their lavish use of precious stones. Henry III. and his queen possessed magnificent jewels and other decorations richly garnished with them. The wedding gifts of this royal bride included nine chaplets for her hair, formed of gold filagree and clusters of gems, great camei brilliant with costly stones, a silver peacock with the train made of pearls, sapphires, and other rare stones, used for sweet waters which flowed from the beak into a silver basin, and eleven garlands enriched with pearls, emeralds, sapphires, and garnets, a present from the bridegroom. The royal crowns and girdles of this princess were all garnished with a profusion of the richest gems the mineral kingdom could afford. Edward III. was very liberal in his gifts, bestowing them upon his favorites, one of whom received nearly twenty thousand specimens of pearls and precious stones. Richard II. expended upon a single coat jewels costing nearly one million dollars.

The Tudor princes were no less conspicuous for their love of personal ornaments than the Plantagenets had been. The coronation robes of Henry VIII. were resplendent with gold and precious stones, and a gay bridegroom was this same prince if we may believe the reporters of his time, who say he was attired in cloth of gold loaded with diamonds, emeralds, rubies, and pearls. His costume, at the celebrated meeting of the "Cloth of Gold," was conspicuous for its richness even where each prince was emulous to outshine every other, in the splendor and magnificence of his wardrobe. His daughters inherited their royal father's tastes, seen in the richness of the bridal trousseau of Mary Tudor, and at the court of Elizabeth, who is said to have surpassed all her contemporary sovereigns in the profusion and variety of her jewels. The nobles and courtiers imitated the example of their queen, so that her reign was signalized for the brilliancy of her court in costly attire, as well as brilliancy in wit. It was the custom of this period to confer valuable jewels as guerdons upon knights of the tournament and prize-fighters; the queen, on a similar occasion, presented to a successful athlete a jewel set with rubies and diamonds, valued at more than three thousand dollars, a gift far more costly in those times than it would be now. The old-fashioned devices and mottoes called "posies," frequently alluded to by contemporary writers, were spelled or represented by precious stones.

Mary Queen of Scots was the peer of her rival in the possession of rich jewels. Her crown worn at her marriage with the Dauphin was brilliant in diamonds and other valuable gems: a single carbuncle pendant was valued at hundreds of thousands of dollars; her train, six yards in length, was so loaded with precious stones that it had to be borne through the mazes of the dance by an attendant. The jewels of this princess named in her will, comprised, with many others, a diamond necklace, ruby chains, and a parure of pearls brought from France, considered the finest in Europe at the time. Some of these treasures were deposited in Edinburgh Castle, others were scattered or stolen after the defeat of her army at Langside.

Anne, wife of James I. of England, was the owner of a large collection of precious stones, nor was the king himself insensible to their attractions, as appears from the display of three million dollars worth of jewels on his person at the marriage of his daughter to the Elector Palatine, and this was before the Koh-i-noor came into the possession of the royal family. All the sovereigns of the Stuart line were opulent in precious stones. Charles I., notwithstanding his poverty and long wars, found means to enlarge his inherited possessions by purchases, which, in a single year and a half, amounted to a quarter of a million dollars. Mary, the wife of James II., wore at her coronation jewels estimated at a sum between one and two millions. On the abdication of this sovereign, nearly all his private jewels, including the ornament of Mary Queen of Scots, which has a tragical history, and some belonging to the crown, were carried out of the kingdom and scattered abroad. Queen Caroline, consort of George II., owned a large collection of gems; her diamonds alone were valued at five million dollars, and yet at her coronation she borrowed jewels of some of the court ladies lest her own should prove insufficient for the occasion. A tiara of diamonds, worth three hundred thousand dollars, adorned the brows of Charlotte, wife of George III., while one lady of the nobility wore, at the crowning of this ruler, diamonds estimated at nearly one million. The costumes of the Duke of Buckingham were always of the richest kind; his cloak, hat, plume, girdle, sword, and spurs, were brilliant with costly diamonds. One of his suits, made of white velvet, was embroidered with nearly half a million worth of these gems, and another, of purple velvet, was covered with valuable pearls. When sent to France to escort Henrietta Maria to England, as the bride of Charles I., this nobleman took with him twenty-seven different suits for his own use, all embroidered or ornamented with precious stones. Officers of state, up to this period, had indulged in the same luxurious habits which characterized royal and noble families. The belt, girdle, and baldrick of the Lord Chancellor, during the reign of William the Conqueror, were remarkable for richness of ornament, and shone "like twinkling stars with stones most precious rare." The "St George" belonging to the Garter, though not now ornamented with precious stones, was formerly garnished with costly gems. The insignia sent to Gustavus Adolphus bore a "St George" enriched by eighty-four large diamonds.

The style of dress among gentlemen of the present day does not admit of an excessive use of jewelry; but it was otherwise with past generations, when the materials for costumes consisted of satins, velvets, and embroidery, and when the fashion of their garments allowed such ornaments. Less than two centuries ago, a gentleman's attire was not complete without "a jewel for his hat, chains for his neck, and rings for his fingers," all more or less enriched with precious stones. These elaborate wardrobes, says Mme. Barrera, proved a heavy burden in a pecuniary sense; and not infrequently public men, especially ambassadors to foreign courts, not only spent all their income to maintain their station, but often incurred heavy debts besides.

A bride's trousseau in the sixth century exceeded even that

of the present day. Fifty wagons were required to convey the wedding presents of one of the princesses of those times, consisting of gold, precious stones, and rich apparel ornamented with gems. The trousseau of Isabella, wife of Richard II., surpassed in splendor anything before seen in England, comprising, with other rare curiosities, a robe and mantle of velvet, embossed with birds composed of different kinds of gems, resting upon branches of emeralds and pearls. It sometimes happened that the weight of her jewels was too burdensome for the fragile form of the bride, as was the case with Jane, a princess of Navarre, who, at her marriage with the Duke of Vendôme, had to be carried to the chapel in the arms of the Constable de Montmorenci. This princess was the proprietor of immense wealth in jewels and precious stones. An apartment in her palace was filled with various articles made of them — cups of agate and rock-crystal studded with sparkling gems, mirrors bordered with diamonds, gold dishes set with rubies, pearls, and turquoises, salt-cellars, vases, and other articles, all garnished with costly ornaments.

Gems were used in embroidery, and hardly an article of dress upon which they could be displayed was deficient in these decorations. Coverings for the feet afforded conspicuous objects for such an exhibition; and this accounts for so many examples of ornamented shoes among the relics of the Middle Ages. Those of Charlemagne, in the Vienna Museum, and of Cardinal Wolsey, which are said to have been worth thousands in gold and precious stones, afford illustrations.

Fans are, perhaps, the most interesting articles of the feminine toilet, for the part they have played in coquetry; and it is not surprising that they have been decorated by the most brilliant and fascinating ornaments nature yields. The fan of

Marie Antoinette was remarkable for richness and curious workmanship. Its gold handle, enamelled and embellished with pearls, rubies, and bouquets of diamonds, supported a frame of ivory, imitating lace-work, mounted with the imperial eagle. This trinket, like so many other royal jewels, has its romance, having been used by two unfortunate queens of France — Marie Antoinette and Eugénie.

Princes and nobles have competed for the monopoly of the trade in precious stones, and frequently their collections assume vast proportions, representing uncounted wealth. According to an inventory of the jewels of a Swedish noble, made at the close of the sixteenth century, his cabinet contained one hundred and eighty-four large diamonds, forty-six rubies, four hundred and sixty-one emeralds, two hundred and fifty-six pearls, besides an unlimited number of mounted gems, or those otherwise disposed of.

Prince Potemkin (1736–1791), one of the most accomplished courtiers of the Empress Catherine II., was the owner of a choice collection of precious stones, including a large number of diamonds, with which he was accustomed to amuse himself by arranging them in various ways on a table covered with black velvet — a harmless, if not a dignified, pastime. The court favorite was fond of displaying his jewels upon his person, and appeared on important occasions arrayed in the splendor of an oriental despot. His collar of one of the Russian orders was sparkling with diamonds worth hundreds of thousands, while millions more were represented in diamonds, emeralds, and other valuable gems, which adorned a wreath and epaulets belonging to this Muscovite prince. Suvoroff, the celebrated field-marshal, stern Cossack as he was, had a great admiration for precious stones, and was the owner of many fine gems, gifts from the different sovereigns of Europe, which he carried with him on his military campaigns, to be studied and admired during his leisure moments.

The Esterhazy gems "have passed into history," and fill many a page with their glittering illustrations. Prince Nicholas, one of this ancient Hungarian family, who flourished at the close of the eighteenth and the beginning of the nineteenth century, revived the mediæval practice of decorating military uniforms and weapons with precious stones. All his armor worn at the coronation of Francis II. as King of Hungary was covered with valuable gems; his sword and scabbard sparkled with brilliants; a shoulder-band and belt presented a galaxy of diamonds and pearls; while his numerous Orders enhanced the splendor of his equipments. The uniform of the prince as a Hungarian general was ornamented with fifty thousand diamonds, besides many fine specimens of rubies, topazes, emeralds, and other varieties of precious stones. The cap was encircled by a band of pearls, and bore a plume composed of five thousand diamonds of different colors. With all his magnificence, Nicholas Esterhazy was excelled in the display of splendid jewelry, by his successor, Prince Paul, who died a bankrupt, in 1866. His large collection was sold at his death, to liquidate his debts, and the celebrated Esterhazy gems were scattered over the world. They included, among others, more than fifty thousand brilliants, an aigrette of diamonds for the prince's military cap, a loop of diamonds and pearls, and a sword, sheath, and belt, most elaborately ornamented with precious stones.

The question naturally arises: Where did all the gems come from? Mr. Jones, in his work on precious stones, says the Crusaders introduced vast quantities into Europe, along with many other luxuries. It is related that one hundred and fifty mules were needed to transport the spoils of Tancred, including gold, silver, and precious stones; while an immense quantity of the same kind of treasure was captured at Cyprus, by Cœur-de-Lion. It may appear incredible that these articles, especially gems, which, from their small size and indestructible nature, could have been easily concealed, should not have eluded their captors; but it will be remembered that much of the spoil was plundered from besieged towns whence escape was improbable, or was captured on the battle-field, in consequence of the custom of princes to take to the scene of war their most valuable jewels, which were often worn as ornaments during an engagement, as was the case at the battle of Grandson.

EQUIPAGE AND ARMOR.

The practice of decorating the trappings of horses, armor, and royal equipages, though not confined to the East, originated there in very early times, as the Assyrian bas-reliefs at Nineveh prove; and this time-honored fashion is still in vogue among some of the nations of the present day. It was a Roman custom, in use before the Empire, as is shown during the civil war between Metellus and Suetonius, in Spain, when Pompey's life was saved in one of the battles by his richly caparisoned horse. This general, being hard-pressed by the foe, dismounted, and, turning the animal, which was covered with decorated trappings, out among the enemy, made his escape while the soldiers were quarrelling for the booty. It is seen by the numerous collections in the museums that the practice was universal throughout Europe during the Middle Ages; at a later period, Charles II., of England, ornamented the stirrups of his saddle with three hundred and twenty diamonds, while the palfrey of Mary Queen of Scots was caparisoned with purple velvet, embroidered with pearls, and a bridle richly set with precious stones. The armor of knights, both offensive and defensive, was similarly embellished.

The Asiatic princes of modern times have preserved this custom of their ancestors, of ornamenting with precious stones almost everything they used, from their diadems to their chariots, and all their household utensils. The royal equipage of Runjeet Singh was covered with them, while his favored horses were honored by being occasionally allowed to wear the Koh-i-noor. The horses of Maha Raja were accoutred in jewelled harnesses, with emeralds of great size suspended from their necks. These equine decorations were valued from one million to one and one-half million dollars. The King of Siam does not regard diamonds, rubies, emeralds, and pearls, too costly for garnishing his horses and carriages — a sentiment shared by his neighbor, the King of Burmah, whose state-carriage, captured by the English, in one of their wars with this country, was ornamented with precious stones estimated worth sixty thousand dollars.

It was stated in the public journals that the Czar of Russia presented a magnificent sword to General Komaroff as a token of his regard, which was made of Damascus steel, with the scabbard and hilt of gold, ornamented with rows of jewels and clusters of large diamonds, thus showing that the ancient European custom of embellishing armor has not become entirely obsolete.

CHAPTER VI.

DIFFERENT KINDS OF ORNAMENTS.

Rings. — The ring is one of the oldest and one of the most common jewels known. Its great antiquity is proved by the statues at Elephanta, India, and by other relics of the past, as well as by the records of the sacred writers. As a symbol, it represents power, honor, rank, and alliance, and was the emblem of authority among the ancients, as with the Egyptians, when Joseph was invested with the vice-royalty of the kingdom, and with the Persians, when Ahasuerus appointed Mordecai to supersede Haman in office, and when, by command of the King of Babylon, it was employed to sanction a proclamation in favor of the Jews. The ring was also used by the Israelites to make valid royal decrees, as when Jezebel, the queen of Ahab, fraudulently employed his ring for this purpose; and to authenticate public documents, and legalize business transactions between contracting parties. In these instances it is probable the ring was mounted with an engraved stone as a seal, since this jewel was frequently set with gems bearing inscriptions, and used for signets by the nations of antiquity.

The ring has been very generally selected for the ceremony of investing persons with sacerdotal prerogatives; for betrothments, marriages, and memorials for the dead; while as a personal ornament, there have been no limits to the numbers worn. Seneca says of his countrymen: "Our fingers are loaded with rings; each joint is adorned with precious stones"; while one prominent Roman, who was, undoubtedly, a representative of many others, displayed sixteen rings on his fingers at once. The Romans had their summer and their winter rings, and the Greeks had their rings for each week in the year. Heliogabalus, who has been called the greatest fop of all the Emperors of Rome, never wore the same ring twice; but this effeminate prince was excelled by a German noble of a later period, who had not only a ring, but, a snuff-box, both ornamented with gems, for each day in the year.

The most important quality of the ring, according to tradition, was its power to endow the possessor with the gift of eloquence; hence, they became important to lawyers when pleading, and to poets when reciting their verses, and they ought to be in great demand in these times of oratorical efforts. When this jewel was set with a stone engaved with representations of the planets, it was thought to be invested with remarkable virtues, and was employed in mystic rites; it became a sure protection against certain maladies, after receiving the royal blessing, and it is still regarded by many as a talisman.

The Gemmel, or Jumelle.—The betrothal ring is very ancient; some antiquaries believe it is of Hebrew origin. Marriage rings set with intagli were in use among the Romans at an early period in their history, as represented in their ruins; and probably the gemmel was familiar to them. In Pliny's time, it consisted of iron set with lodestone, signifying a mutual surrender of liberty; sometimes it was made of two hoops, one inside the other, engraved with the names of the betrothed, and could be worn separately. The belief, still prevalent, that the loss of the bridal ring is an ill omen, is a superstition which had its origin in a past age.

The ring was used at the imposing ceremony of espousing the Adriatic inaugurated at the defeat of Frederick Barbarossa by Admiral Ziani. Pope Alexander presented this jewel to the doge, saying, "Take this ring and with it the sea as your subject. The annual return of this day shall commemorate the subjugation of the Adriatic to Venice as a spouse to her husband." For six hundred years the ceremony was yearly observed by the doge's letting fall into the sea a ring, pronouncing at the same time the formula: "We wed thee with this ring in token of our own true and perpetual sovereignty." For more than fifteen centuries, this emblem has performed an important rôle in the ceremony of consecrating a bishop. The ring used for this purpose is made of gold mounted with a ruby, sapphire, or amethyst, more frequently the latter, from its supposed moral power over the candidate. The Roman pontiff has two rings, one for his special use, and another for that of persons acting under his authority, as well as for his own accommodation: those of very large size are supposed to have been employed as credentials for envoys, and not the rings of investiture. It was an ancient practice to bury the ring and other ecclesiastical insignia with the deceased prelate, a custom which proved to be a temptation to frequent robberies of tombs, until imitations were substituted for the real jewels. A large ring of gilt bronze, set with amethysts, was taken from the tomb of Pope Boniface, during the insurrection at Rome in 1849.

The ring used in the coronation of the sovereigns of Great Britain consists of a plain gold band with a large ruby engraved with the Cross of St. George. The one sent to Elizabeth to notify this princess of the death of Mary Tudor, and her accession to the throne, by a strange coincidence, was the same little messenger despatched on a similar errand to James VI., of Scotland, on the death of *his* predecessor.

Rings designed for finger ornaments were not only made of gold and precious stones, but also of silver and the baser metals —bronze, iron, and lead, frequently set with engraved stones; sometimes these jewels were contrived with a hollow cavity for holding poison as a convenient and ready means of suicide, a use to which they were applied by some celebrated characters in history, as is related of Hannibal and Demosthenes. They are assigned an indefinite age by tradition, which represents Prometheus wearing a ring mounted with a stone from the Caucasus; they were very generally used for ornaments, set with intagli, during the Roman Republic. It is said they were collected by bushels from the field after the battle of Cannæ, which occurred in the Second Punic War, 216 B. C., showing that the Romans and Carthaginians had been accustomed to a free use of personal ornaments long before the period of the Roman Empire.

There are historical rings famous for the part they have played in human affairs, or for some striking peculiarity they have possessed; one of the latter class was the agate ring of King Pyrrhus, said to have presented a natural representation of Apollo and the Muses; and of the former, the ring given by Ptolemy to Lucullus when sent by his government on a mission to Africa, containing a valuable emerald engraved with the king's portrait, which the Roman consul accepted with great reluctance lest he might be accused of having been bribed. The ring of Polycrates of Samos is one of the most celebrated jewels of antiquity. The story, briefly told, is as follows: This prince, having been favored with uninterrupted good fortune, was advised by his ally, Amasis, King of Egypt, to make some costly sacrifice in order to neutralize the danger of unalloyed prosperity. With this object in view he threw into the sea his signet ring set with an emerald, his most valued gem. But

Fortune, as if determined upon his destruction, ordained that the jewel should be swallowed by a fish, which was soon after caught and brought to the monarch's kitchen; consequently the ring was restored to him. His royal confederate, fearing he himself might become involved in the destruction sure to overtake Polycrates, withdrew from the league — just in time to secure his own safety, for, soon after, the Samian king was taken captive by the satrap of Sardis and crucified.

The story has, undoubtedly, some foundation in fact, since Herodotus refers to the emerald ring of Polycrates, engraved by Theodorus of Samos. This jewel has given rise to some speculation about its identity; the one deposited in the Temple of Concord at Rome was claimed to be the famous ring, but that was made of sardonyx and not engraved, which does not answer to the description of the Greek historian. An emerald found in Aricia, Italy, a few years ago, of large size, and engraved with the figures of a lyre and cicada, an insect known for its musical powers, has been thought to be the identical emerald of Polycrates, but it is highly improbable that a gem dating back more than twenty-three centuries, should ever be recovered and identified.

The Dresden Museum contains more than sixty rings, many of them historical, set with different gems — diamonds, rubies, emeralds, opals, sapphires, hyacinths, garnets, carnelians, and other precious stones. One of the collection, with a sapphire d'eau, was given to a knight by the Elector John Frederick when the latter was taken prisoner at the battle of Mühlberg, 1547; two others formerly belonged to Martin Luther, and one to Melanchthon.

During the mediæval period of art, rings were set with gems engraved with classical subjects, made to represent Scripture characters; as Isis personated the Virgin; Jupiter, the Apostle

John; Serapis denoted our Lord; and Cupids were used for cherubs. The word bagues (rings) was formerly synonymous with personal effects or baggage; hence, in capitulations, the phrase, "sortie vie et bagues sauves," meant to depart with life and rings safe — that is, with all one's personal property. Perhaps the origin of the English phrase, "bag and baggage," may be traced to the same source.

Ear-rings. — Jewels for the ears as well as the fingers have a great antiquity, and have been worn in all ages for decoration, and sometimes for amulets and talismans. The Persians and Peruvians have always cherished a special fondness for this ornament, and frequently wore them of large size and remarkable beauty. They constituted an important article in the regalia of their kings and rulers, as may be learned from the numerous ear-rings found in the tomb of Cyrus, at Pasargadæ, and from the narratives of the Spanish conquerors of Peru. These jewels were sometimes made to be inserted in an orifice pierced in the lobes of the ear, which was enlarged from time to time, until of the necessary size, instead of pending from that organ, after the usual custom. They have in some instances been regarded as the badge of servitude, as well as the symbol of rank. The Jewish rabbies say that Eve was condemned to have her ears bored after her expulsion from Eden, as a sign of her subjection to her husband. If that is true, she had her revenge, in the great expense it entailed upon him to supply the necessary jewels. That they were worn by persons of both sexes and of all ages in ancient times is shown from the great number of these ornaments among the Israelites, who bestowed them liberally to make the golden calf and the ephod of Gideon, unfortunate offerings in both instances. After his defeat of the Midianites, this valiant warrior obtained as spoils a collection of ear-rings, which altogether weighed one thousand seven hundred shekels of gold. Job's friends, in the days of his returning prosperity, gave him an ear-ring apiece, implying they were to be worn singly, or as mismated pairs.

Homer, who understood gods as well as men, adorns Juno with ear-jewels, to captivate her inconstant spouse; hence we infer their use was familiar to the Greeks in his time. Pearls were the favorite gem for ear-ornaments, both with the Greek and Roman ladies, though various kinds of precious stones were employed for this purpose. The fashion was carried to such an excess by the women of the Roman Empire that the guardians of the public welfare felt constrained to condemn their luxurious habits in the most unequivocal language.

These jewels were at one time worn in England in the form of keys, a custom alluded to by Shakspeare.

Bracelets. — Both armlets and bracelets have been considered a necessary appendage of royalty, especially by orientalists; while in ancient Rome they were the symbols of honor or the badge of servitude, according to the material of which they were made, whether gold or iron. The bracelet, like the ear and finger ring, can claim a high antiquity. The one presented to Rebekah weighed one shekel, the value of which has been variously estimated; and those worn by the Sabines were of such beauty and richness as to ensnare the unfortunate Tarpeia into betraying her country, and, as a penalty for her crime, she has ever since, according to legend, been compelled to sit spellbound upon the Tarpeian Rock, in Rome, covered with the jewels she so much coveted.

The Assyrian kings are represented in the bas-reliefs of Nineveh adorned with bracelets; and these ornaments have been found in Egyptian remains inlaid with precious stones. The use of jewels by the ancient Egyptians must have been very general, since they bestowed them with a liberal hand upon the Israelites, to expedite their departure from the land of their terror-stricken oppressors. The account of the sacred historian is confirmed by the paintings at Thebes representing the scene.

The ancient Britons were accustomed to adorn their persons with bracelets, according to the poems of Beowulf, and descriptions of the costume of Queen Boadicea. The Anglo-Saxons adopted the fashion, which has descended to their successors down to the present day, making the bracelet an indispensable article to a complete toilet. Though generally worn upon the wrist, it was sometimes used as an armlet; the Emperor Maximilian, who was of gigantic size, is said to have worn the bracelets of the empress upon his thumbs. The richest jewels of this kind belong to the Persian regalia; those found among Chaldæan and Assyrian ruins were made of a bar of metal tapering at one end, and bent into an oval, and, as art advanced, precious stones were used for their embellishment.

Necklace. — Homer alludes to the gold and amber necklace presented to Penelope by one of her suitors; a reference to this ornament, and the story of the fatal necklace of Harmonia, prove that the early Greeks were acquainted with its use. Like the bracelet, it was made of gold, silver, or a series of gems strung together; and sometimes three of these jewels were worn at once, with a girdle of the same kind of precious stones around the waist, from which was suspended a bottle of perfume. In the fifteenth century, the necklace had expanded into a broad collar thickly set with precious stones, with a pendant attached; and at a later period, they were made to be worn looped to the girdle or grouped about the shoulders. The queen of James I. of England was accustomed to wear several of these ornaments at the same time, with a baldrick or garland hanging from the shoulders to the waist.

Tavernier describes the women of Tonquin as being extravagantly fond of adorning their persons with necklaces made of coral and yellow amber; those worn by the Hindoo princes were mines of costly gems. A magnificent specimen, captured by the Sultan Mahmoud, composed of large pearls, rubies, and other costly materials, was valued at half a million of dollars.

The story of Serena, wife of Stilicho, a prominent Roman of the fifth century, affords a tragical illustration of the fascination which precious stones have over some minds, and the swift retribution of a sacrilegious act under the Roman laws. A necklace of costly gems placed as a sacred offering upon the statue of Vesta was coveted by this high-born lady, who finally appropriated the tempting jewel, for which act of impiety she was condemned by the Emperor Honorius to suffer death by strangling, notwithstanding her exalted rank. The most remarkable necklace of antiquity was that of Harmonia, who received it, together with the famous peplum, as a wedding gift on the occasion of the marriage of this goddess to Cadmus. It proved a fatal jewel to every mortal who was so unfortunate as to possess it. Polynices, who inherited the necklace, gave it to Eriphyle as an inducement to use her influence with Amphiaraus, her husband, to join the expedition against Thebes. He yielded to her entreaties and perished in the enterprise, according to his own prediction: Alcmæon, their son, killed his mother and possessed himself of the ornament, but he came to a tragical end, and, after changing owners several times, to all of whom it proved disastrous, it was dedicated to the gods in the Temple at Delphi. The description of this necklace is given by Nonnus, a poet of the sixth century, who says it was made in the form of a serpent, with two heads bearing a golden eagle with four wings; one pinion was of orange jasper, another of white moonstone, a third was made of pearl, and the fourth of Indian agate. A ruby reflected its lamp-like flame from the heads of the serpent, while the eyes where composed of the lychnis, a fiery-red stone, and the pendant consisted of an emerald and a crystal surrounded by a setting representing birds and fishes. Mr. King says this necklace is the most ancient jewel on record, and thinks it was not altogether the creation of the poet's fancy, but the exaggerated reproduction of some very ancient relic of prehistoric times. He suggests the idea that the Greeks may have borrowed the legend from the Assyrians, as the serpent and the four-winged eagle were Assyrian types, unknown to the Greeks.

Signets or Seals. — The origin of *sigilla*, or seals, has been traced to the Israelites — it is supposed, on account of the numerous Hebrew words and titles of the Deity which occur on Gnostic intagli, but it is quite likely they were employed by the Egyptians and Chaldæans long before the birth of the Hebrew nation, since they have been found in the remains of a very remote antiquity. They were of different kinds: sometimes a single gem, set expressly for the purpose, and engraved with different emblems, a motto, or the owner's name, answered for a signet, designed to be suspended from the neck or arm; at other times, they constituted the stone of a finger-ring, and with the Chaldæans, they assumed the form of cylinders. "Every Babylonian," says Herodotus, "had a seal." How this instrument came into use has been told by Perrot, somewhat in the following manner: At first the jewel-boxes were supplied with pebbles from the beds of rivers, which were drilled through the centre to be worn for ear-rings, bracelets, and necklaces, many of these primitive ornaments having been found in Chaldæan and Assyrian tombs. But the artists of those times were not long contented with these simple jewels, and the fancy to engrave some design upon them resulted in a more finished style of work, so, from holding the mere rank of ornaments, these engraved gems became the legal instruments of authoritative documents, and the *seal* came into being. To invest it with magical powers, the figure of a god was inscribed upon it, and thus the seal answered the double purpose of giving sanction to contracts, and serving as a talisman to the owner in virtue of the representation of the deity it bore. The subjects for these engravings varied, but the favorite one was the triumph of the gods over demons.

The cylinders used for signets, as well as personal ornaments, by the Babylonians and Assyrians were, as a rule, from three-fourths to two inches in length, with a surface generally convex, sometimes concave, engraved with intagli, and drilled through the centre lengthwise, for the purpose of attaching them to the person. They were mostly cut in lapis-lazuli, jasper, and other quartz gems, and sometimes hematite. The subjects selected for these engraved seals comprised both men and animals, representing combats, and sacrifices to the gods. The manner of taking an impression with these cylinders was by rolling them over a lump of tempered clay laid upon the object to be sealed, a practice to which Job alludes, when he compares the heavens bristling with stars to these impressions. He says, "It is turned as clay to the seal." Cylinders are found in great numbers in the ruins of Chaldæan and Assyrian cities. The Persians made use of conical or spheroidal blocks of precious stones — chalcedony and agate being employed most frequently — for their signets, as well as the Babylonian cylinder.

It is thought the early Greeks were unacquainted with the use of signets, as Homer makes no allusion to them; but they were in vogue among the Romans, to make valid wills and other writings, as they are at the present day. They adopted the Etruscan and Egyptian use of the scarab for this purpose, a practice which was retained until late in the Republic. The ancient Mexicans and Peruvians had their seals, an evidence that all civilized nations considered them essential in the transaction of business.

The seal formerly used by distinguished persons corresponded to the coat-of-arms now employed to designate eminent families. Some of these emblems have become historical through the fame of their owners; such are the signet of Darius, made of green chalcedony, now in the British Museum; the one supposed to have belonged to Sennacherib, of Amazon stone; and the seal of Michael Angelo, in the Paris collection. Alexander the Great made use of two different signets — that of Darius, in his edicts to the Persians, and his original seal, engraved with a figure of the lion, for the Greeks. The head of Augustus, engraved by Dioscorides, was used by the Roman emperors, until Galba substituted his own seal, which bore the figure of a dog looking from the prow of a ship. That of Sylla represented the surrender of Jugurtha; that of Pompey, a lion holding a sword in his paw; that of Augustus, the figure of a sphinx, afterwards changed to his own portrait, in consequence of the sarcasm it called forth; while Mæcenas adopted the frog as his emblem. Documents bearing his seal inspired the Romans with terror, who regarded them as signals for a new levy of taxes. The signet of Charlemagne was engraved with the figure of Serapis, an Egyptian divinity — a singular preference for a valiant Christian prince.

The most famous seal of later times is the one used by Michael Angelo, made of sard, engraved with a group representing a Bacchic festival, in the Renaissance style, though some virtuosos believe it is a genuine antique. A curious story is told how this seal was surreptitiously swallowed by a distinguished antiquary, when on a visit to the Bibliothèque, in

Paris, and how an emetic was promptly administered, by which the jewel was recovered. It has been stated there are more paste copies of this seal than of any other engraved gem, doubtless on account of the celebrity of its original owner. Two different seals are used by the Roman Pontiff; one consists of a large ring, the annulus piscatores bearing the effigy of St. Peter drawing a net, used for briefs and private letters; the other, employed for bulls, has on one side the heads of Peter and Paul, with the cross between them, and on the other the portrait of the reigning pontiff, and sometimes his coat of arms. At the death of the pope, his seals are broken, and new ones are given to his successor by the city of Rome.

Brooches were worn by Etruscans, Romans, and Saxons, for talismans, and have been used since for ornaments. The "Brooch of Lorn" was "burning gold, studded fair with gems of price."

Diamond buttons, as fastenings for the dress, were used by both sexes during the sixteenth century; but this custom, as a prevailing fashion, has fallen into desuetude.

Chains. — These ornaments, embellished with all manner of precious stones, have a great antiquity, and constituted an indispensable part of the attire of a courtier or public functionary. The chain was a symbol of investiture used by Pharaoh, King of Egypt, and many centuries later by Belshazzar, King of Babylon, when Joseph and Daniel — both Hebrews — were advanced to position and power in the government. The chain has ever since been considered an important jewel, either as an emblem of rank or authority, or as an ornament, until, in modern times, it has become an essential adjunct to the watch — an invention of a comparatively recent period.

CHAPTER VII.

SACRED USES OF PRECIOUS STONES.

PAGAN and Mohammedan, Jew and Christian, have given alike, with unstinted generosity, the choicest of earth's treasures to embellish their temples, mosques, churches, sacred vestments, emblems, and utensils; hence we find that precious stones have been consecrated to the purposes of religious worship from a remote period down to the present time. Nor have they been withheld from tombs, shrines, and other memorials for the dead which loving hands have reared to departed friends. Among heathen nations, the most beautiful and valuable things were dedicated to their divinities. Whatever was most rare and costly of the spoils taken in war, and whatever was most magnificent and elegant in art, were devoted to propitiate their favor, appease their anger, or as tokens of gratitude for blessings enjoyed. This was especially true in the Roman Empire after her numerous conquests, when the most costly gems collected from the different countries subjugated by her powerful armies were poured into the sacred treasury at Rome, as free-will offerings.

Both Pompey and Cæsar presented some of their richest spoils to the temples of their favorite gods. Pompey consecrated the treasures captured from Mithridates, comprising rubies, topazes, emeralds, opals, diamonds, and stones of inferior rank, besides numerous rings, bracelets, and gold chains, of exquisite workmanship. Cæsar devoted six caskets of his choicest engraved gems to Venus, and Alexander Severus dedicated to the same favored goddess gifts of pearls of remarkable size and beauty, which had been presented to him by the Persian envoy to Rome. Augustus and many other distinguished Romans gave freely, to maintain their national religion, vast stores of gold and precious stones.

The statues of the pagan divinities, as well as their temples and shrines, were frequently adorned with these costly offerings, that of Jupiter Olympus being a notable example. The Syrian goddess Astarte was honored by munificent gifts in precious stones from her numerous worshippers, who visited her shrine from all the nations of the world.

These symbols of idolatry, the images of the gods, sometimes served a double purpose — that of awakening the religious emotions of the laity, and securing generous offerings, and at the same time affording a secret hiding-place for these gifts, appropriated by the sacerdotal class. When Mahmoud wrested the heathen temples of India from the Brahmins, there was found inside the celebrated statue Summat erected in Guzerat, a vast accumulation of pearls and different kinds of precious stones, the donations to the god made by his unsuspecting worshippers, which the crafty priests had purloined and concealed in the capacious stomach of the huge idol.

Gems were frequently devoted to sacred uses by having the head of some divinity engraved upon them, a custom prevalent in Egypt, as we learn from gems cut in amethyst in the form of a pyramid bearing the figure of the god Serapis, an Egyptian deity, though the temples and sacred relics of this country afford fewer examples of the use of precious stones than the ancient shrines and pagodas of India.

Precious Stones of Paradise. — All religious beliefs have included gems in the decorations of the future abode of the righteous. In Lucian's "Island of the Blessed," the walls of the city were built of emeralds, the temples of beryl, the altars of amethysts, and the mansions were of gold. The paradise of the Chinaman is adorned with gold and precious stones; that of the Moslem is paved with pearls and jacinths, its rivers flow over pebbles of rubies and emeralds, and its happy residents, conspicuous for their crowns and bracelets of gold and gems, regale themselves in tents of pearls, emeralds, and jacinths. The ideal of the Holy City presented to the mind of a Jew is described in the prayer of Tobias, who predicts that "she shall be built up of emeralds, sapphires, and all precious stones, her walls and battlements of fine gold, and the streets shall be paved with carbuncle, beryl, and stones of Ophir."

The New Jerusalem of the Christian dispensation exceeds the powers of the imagination to conceive the splendor of its walls of precious stones, its streets of shining gold and crystal, than which nothing can be more magnificent. Mr. King suggests that the author of the Apocalypse must have been familiar with the character of gems and the effect of their color when arranged by the side of one another; but it is probable that the ancient names do not correspond in all cases to the modern.

Let us imagine, if possible, the exterior splendor of this city, admitting, for the purpose, that we understand what stone is meant by the name. For the foundation, the dark green jasper is used, supporting a layer composed of sapphire thought to be lapis-lazuli; and above that, the chalcedon, a greenish blue emerald. The fourth stone of the series is smaragdus, a transparent, bright green emerald; while above this rests the red and white sardonyx, superimposed by the bright red sard: this completes one-half of the height of the wall. The seventh gem in the structure is the golden chrysolite; then comes the bluish green beryl, above which is laid a yellowish green topazion, thought to be either peridot or topaz. The tenth stone is the chrysoprase, of dark green hue, surmounted by the jacinth, of rich crimson, and crowned by the purple amethyst.

The Hebrews employed precious stones to decorate the sacerdotal robes, the Tabernacle, and, later, the Temple at Jerusalem. King thinks it probable that the most ancient of all authentic sacred jewels were those of the breastplate of the Jewish high-priest, supposed to be the "Urim and Thummim," though what this was has been a disputed question. The words have been variously interpreted, "Lights and Perfections," "The Declaration," "The Truth," and "The Oracle of Judgment." Some commentators have maintained the opinion that the Urim and Thummim was distinct from the breastplate, and consisted of a blue sapphire worn over it when the high-priest entered the "Holy of Holies"; but the Jewish writers probably understood better than any others the nature of their own sacred symbols,—Josephus applies the mystic words to the breastplate itself.

This priestly ornament was a square of eight inches set with twelve different gems engraved with the names of the tribes of Israel, a stone for each tribe, arranged in four rows. Both Josephus and the Vulgate give a different order from our version, but the stones are the same except the chrysolite substituted for the diamond. They are as follows:—First row: sard, red; topaz, yellowish green; smaragdus (emerald), bright green. Second row: carbuncle, red; sapphire, blue; jasper, green. Third row: ligure, yellow; achates (agate), black and white; amethyst, purple. Fourth row: chrysolite, yellow; onyx, blue and black; beryl, pale green or blue.

Josephus says these stones shot forth brilliant rays of fire to denote the presence of the Deity, but this power ceased two hundred years before his day, in consequence of the sins of his people.

The first breastplate made for Aaron was lost during the Babylonian captivity; therefore, after the restoration of the Jews, it was necessary to replace it by a new one, an exact copy of the original except the inscriptions, which in the first were in the Hebrew language, but in the last in Chaldee, or Syro-Chaldee. The second breastplate fell into the hands of the Romans at the capture of Jerusalem, and, after being exhibited with other sacred treasures in the triumphal procession, it was deposited in the Temple of Concord at Rome. When the Empire was overthrown, these sacred trophies were scattered among the victors, and their authentic history ends with this event. Several hypotheses have been put forward respecting their final disposition. One is that they were returned to Jerusalem and were captured by the Persians in the seventh century; another account relates that they were carried to Babylon, then back to Jerusalem, then to Rome, from there to Carthage, thence to Constantinople, then back again to the Jewish capital, and finally to Persia, and may possibly be found in the royal treasury at Teheran. What a remarkable history these memorials have had — fit types of the nation to which they belonged.

An interesting subject for speculation has been suggested regarding the gems of the first breastplate, which is that they may have been reset for the Babylonian or Assyrian kings, as they were of large size and great intrinsic value, as well as objects of interest, and may possibly be recovered, an event no more remarkable than the preservation of the seals now in existence bearing the name of Thothmes III., supposed to be contemporary with Moses.

Some idea can be formed of the richness of the furnishings

for the Tabernacle by the abundance of the voluntary offerings of bracelets, ear-rings, tablets or pendants, and other jewels, presented by the Israelites for this purpose during their exile in the wilderness.

The figure twelve appears to have been a favorite symbol with the ancients, who represented different objects or events by twelve different precious stones — as the tribes of Israel, the months of the year, and other subjects divided into the same number of parts. Twelve gems engraved with anagrams of the name Jehovah, were employed by the Cabalists to predict zodiacal signs, and, under the Christian dispensation, different species were used to designate the twelve apostles, the Christian virtues, and other religious ideas. The diamond symbolized life, joy, and innocence; the ruby, divine power and love; the sapphire, heaven, virtue, truth, and constancy; the emerald, hope, faith, and victory; the amethyst, suffering, sorrow, love, peace, humility, purity, and modesty; the topaz, the goodness of God; the carbuncle, our Lord's Passion.

The official vestments of the prelates in the Christian churches of the Middle Ages often represented fortunes of costly gems, frequently the gifts of devout princes and nobles. Isabella, queen of Edward II., presented to Pope John a cope embroidered with pearls; Henry III. gave a mitre to the Bishop of Hereford enriched with gems valued at several thousand dollars, and other similar donations were made by persons of rank to the higher ecclesiastics of the Roman Church. Shrines, reliquaries, crosses, vases, and other articles employed in religious services, were more or less embellished with rich ornaments of precious stones. A gold chalice garnished with them, found in 1846, near Chalons-sur-Saône, now deposited in the Museum of Antiquities, at Paris, affords an illustration of this practice of the mediæval period of art.

Churches and Other Religious Buildings.—Pagan temples, with all their opulence in works of art and precious jewels, were equalled, if not surpassed, by the churches and convents of mediæval times in munificent donations and rich ornaments. Princes, nobles, and prelates, vied with one another in their voluntary offerings to these institutions. Constantine the Great signalized his reign by his generous gifts both to Roman and Byzantine churches—an example followed by his successors. King Dagobert of France, in the seventh century, and Charlemagne in the ninth, liberally endowed many of the ecclesiastical institutions of their times with costly treasures in jewels and precious stones. Mme. Barrera says that vast stores of gems and other valuables were poured into the churches and monasteries during the period succeeding the reign of Charlemagne, in consequence of a prevailing belief that the end of the world was near at hand; therefore, these luxuries would be no longer needed by the owners. How the churches, in such an event, should find a use for them, is not apparent; but, having once passed into their possession, there was no possibility of recalling them. The Church of St. Denis seems to have been a favorite among princes, since so many of them contributed to fill its treasury with a vast accumulation of consecrated jewels. The abbey connected with this church contained one of the richest collections of gems gathered from different sources—from the cabinets of the early kings, from conquests, and from foreign princes, as offerings to the patron saint. The church, named for St. Denis, who, according to tradition, was buried here, is about six miles north of Paris, and subsequently became the mausoleum of thirty-five kings and nineteen queens of France, whose combined reigns extended from the time of Dagobert to the eighteenth century. Here were deposited the ring and staff of the saint, the sceptre and sword of Dagobert, together with the gold eagle forming the clasp to his mantle, all garnished with sapphires, rubies, emeralds, and other precious stones, besides numerous shrines, crosses, and chalices, similarly ornamented, and the famous agate vessel known as "Ptolemy's drinking-cup," called, also, Abbot Suger's Chalice, still in existence. The other treasures were lost during some of the intestine wars which have so often distracted this country. Louis VI. and his minister, the Abbot Suger, endowed this church with heavy plate, resplendent with enamel and precious stones, besides other gifts, including a famous crucifix, upon which six or seven artists were employed in its decoration for two years, the gems having been donated for that purpose by different monasteries and other religious houses. The crucifix disappeared, and is supposed to have been appropriated by the Leaguers, during the last of the sixteenth century. Louis VII. followed the example of his predecessor in his pious offerings, which comprised various articles ornamented with antique gems, several of which have escaped the destruction of their less favored contemporaries.

Spain was once very opulent in ecclesiastical jewels; but, like those of France, they have, to a great extent, been scattered or lost during her civil and foreign wars. The Cathedral of Seville and the Church of the Escurial were celebrated for their immense treasures in jewels; while the Cathedral of Toledo surpassed even Sainte Chapelle, at Paris, in the richness and splendor of its shrines, covered all over with precious stones. The figure of the Virgin seated on a rock overspread with jewels, wore on high festivals a gold crown radiant with enamel and gems, surpassing in splendor every royal diadem in the world. The top of this crown was adorned with a superb emerald, which was seized by Marshal Junot, to whom the treasures of the cathedral were shown. The sacred offerings of the Cathedral of Saragossa, though exceedingly rich in pearls and precious stones, were more fortunate in escaping the dispersion which befell the treasures of other Spanish churches during the French invasion of 1809.

The Church of the Virgin del Pilar, a superb edifice, was exceedingly affluent in precious stones. South Kensington Museum contains a collection of jewels purchased from its treasury, including more than five hundred, all ornamented with diamonds, pearls, rubies, and emeralds. The Cathedral of Naples is the repository of the crown of St. Januarius, the patron of the city, which is embellished with three thousand six hundred and ninety gems; and the Certosa of San Martino, in the same city, is a marvel in the profusion of its gem-decorations. The Wenzel Chapel of the Cathedral of Prague is inlaid with Bohemian precious stones; and a very large gilt crucifix studded with gems is seen in the chapel of the old royal palace at Berlin. The cross of King Lotharius, a work of the Carlovingian period (751–987), preserved in the Cathedral of Aix-la-Chapelle, is ornamented with arabesque tracery of pearls, rubies, sapphires, emeralds, and amethysts. At the intersection of the arms of the cross is placed a cameo in onyx, three inches by two and one-half, representing the bust of Augustus. The king's signet on rock-crystal bears the date of A. D. 823.

The Certosa di Pavia, the churches of San Ambrogio, Milan, Or San Michele, Florence, and Santa Maria Maggiore, Rome, are only a few of the numerous instances of the use of precious stones for the decoration of religious buildings. The high altar of San Ambrogio still retains its antique reliefs on silver and gold embellished with gems, — a work of the eighth century. A cross, with several crystal vases, — the work of

Valerio Vicentino, — presented to the Church of San Lorenzo, are seen at Florence. The old Cathedral of St. Paul, London, contained offerings of costly gifts, in the form of reliquaries studded with gems, and shrines covered with gold and precious stones; while Croyland Abbey became the repository of many royal gifts, comprising a globe covered with gems of "dazzling lustre," from the King of France, aud an altar-cloth embroidered with pearls, the offering of Pope Leo IV.

The Roman churches and monasteries were far surpassed in costly endowments by those of the Greek Church, whose numerous chapels, convents, and other religious houses on Mount Athos afford an example of the opulence of this sect in church decorations. One of the monasteries on this mount claims the honor of possessing the girdle of the Virgin ornamented with diamonds and pearls; another contains the veritable cross, set with diamonds and emeralds of remarkable size; a third cherishes with great veneration this emblem of the Christian faith, garnished with diamonds only; while a fourth is endowed with two magnificent crosses covered with gems.

The churches, monasteries, and other religious buildings of the Russian Empire are all profusely decorated with precious stones. The centre of every door in the Hall of St. Elizabeth, in Moscow, says Bayard Taylor, is ornamented with a Maltese cross of large diamonds; and the Cathedral of the Archangel Michael, in the same city, writes Mr. Hamlin, contains ancient reliquaries enriched with a profusion of splendid gems, including a large number of magnificent emeralds and diamonds, while the sacerdotal robes are loaded with jewels of the costliest nature. The patriarchal mitre is all ablaze with brilliant diamonds, rubies, sapphires, emeralds, and pearls, constituting a diadem exceeding five pounds in weight. The convents are, many of them, mines of precious stones, and depositories of valuable jewels, largely the offerings of princes and nobles. In the Convent of Troitza are seen Bibles and other religious volumes bound in covers of silver-gilt overspread with gems and fastened with clasps of antique camei; golden chalices, decorated with rows of diamonds; crosses bearing their rich burdens of emeralds and rubies; dalmatics embroidered with gems; and saints and Madonnas resplendent with brilliant jewels. Two gold crosses, one weighing seventy-five and the other one hundred pounds, both enriched with costly gems, were displayed above the altar in the Cathedral of Constantinople, now the Mosque of St. Sophia, a church under the Christian emperors exceedingly opulent in every kind of decoration, including gold, silver, pearls, and nearly every variety of precious stones.

Memorials for the Dead.—It has been a custom from an early period in the history of the race, to honor the dead with costly tombs or other monuments often embellished with the most valuable objects wealth or devoted love could bestow. For this purpose precious stones, the most coveted of all earth's treasures, have been generously employed not only upon these testimonials of the pious zeal, ardent attachment, or ostentatious vanity of surviving friends, but also upon the mortuary habiliments of the deceased, as Egyptian mummies and the tombs of the saints of the Middle Ages bear witness. Not unfrequently offerings of gems and other precious things were made at the graves of the lost ones, as tokens of affectionate remembrance; especially was this true of the ancient Romans. It is related that Fabia Fabiana, a private Roman lady, dedicated to her deceased granddaughter a large number of valuable gems comprising diamonds, emeralds, rubies, pearls, and other varieties, all mounted in necklaces, ear-drops, bracelets, rings, anklets, and shoe ornaments, as was learned from an inscription on a pedestal supposed to belong to the statue of Isis, found at Alicante, Spain.

The tomb of St. Charles Borromeo, at Milan, is remarkable for the richness of material and the skill in workmanship displayed in its construction. Its columns of choice marble are crowned with gold capitals and draped with crimson damask hangings embroidered with gold. The coffin, made of rock-crystal, and ornamented with the same precious metal, encloses the remains of the saint wrapped in sumptuous robes of the richest fabrics, and in his skeleton hands he holds a crosier embellished with gems and surmounted with a gold mitre. Above the coffin is suspended a crown of precious stones, while the history of the deceased is delineated in bas-reliefs of solid silver. The tomb of Charlemagne was a store-house of jewels and plate. At his canonization in 1166, between three and four centuries after his death, his remains were placed in a golden chair during the ceremony, after having been arrayed in imperial robes and diadem, girded with a jewelled sword, and holding a sceptre in one hand and a gold shield in the other, both garnished with precious stones. These insignia of the distinguished saint were entombed with the remains, but most of them were appropriated subsequently by Frederick Barbarossa. Ivan IV., Czar of Russia, ordered two hundred and fifty thousand dollars worth of gems to be buried with the body of his own son, whom he had murdered.

The Moslems not unfrequently offered the costliest treasures gathered from the different countries of the globe, to their departed friends. The Tay Mahal or mausoleum, built by Shah Jehan at Agra, in the seventeenth century, for his favorite wife, which has been often described by writers and travellers, was the most magnificent in India, and, probably, in the world. It required the labor of twenty thousand workmen for seventeen years to complete it, and contributions of precious stones from every part of his extensive empire to adorn it: — "Jasper from the Punjaub, carnelians from Broach, turquoise from Thibet, agates from Yemen, lapis-lazuli from Ceylon, coral from Arabia, garnets from Bundelcund, diamonds from Punnah, rock-crystal from Malwar, onyx from Persia, chalcedony from Asia Minor, and sapphires from Colombo." Garlands composed of gem flowers, borders consisting of precious stones in imitation of natural vegetation, the most delicate work inlaid with mosaics of these valuable productions, were all displayed upon this world-renowned mausoleum.

Shrines. — These receptacles of venerated relics, cherished both by pagan and Christian, were often garnished with the precious metals, combined with every species of gems the decorator could make use of. The shrine of the Syrian goddess Astarte at Heliopolis became one of the most celebrated of antiquity, and the most costly ornaments were employed for its decoration. The worship of this divinity was celebrated by devotees from nearly every nation of the globe, while her sacrifices were so numerous as to require the constant services of three hundred priests: —

> "With these in troops
> Came Ashtoreth, whom the Phœnicians call'd
> Astarte, queen of Heav'n with crescent horns;
> To whose bright image nightly by the moon
> Sidonian virgins paid their vows and songs."

Portable shrines, often miniature copies of some famous temple with the image of the idol, were common among heathen nations, as the silver shrines of the Temple of Diana at Ephesus, about which such an uproar was made in the time of

Paul. The richest shrine in existence, it has been said, is found in a great temple on an island near the coast of India, dedicated to the god Vishnu, who is sometimes called the Indian Apollo. The jewels consecrated to him are of priceless value, embracing crowns, breastplates, armlets, necklaces, and other ornaments, set with diamonds, rubies, sapphires, emeralds, topazes, opals, and pearls. A single necklace is computed at three hundred thousand dollars, and a net to cover the god's umbrella is interwoven with a vast number of colored pearls, — one hundred and twenty-five thousand, so it is stated. Mediæval Christians, in imitation of pagan customs, carried small shrines covered with jewels in their religious processions, a practice existing in some countries at the present day. Many of the shrines of Italy and Spain, and some of France, have escaped destruction, and are to be seen in many of the museums of curious relics. The shrine of St. Denis, one of the most celebrated in Europe, was honored by gifts from distinguished princes and nobles both native and foreign; the Dukes of Burgundy and Orleans presented at different times munificent offerings in precious stones, thus making it one of the richest, as well as the most famous, in Christendom. The shrine of the Three Kings of Cologne,— Gaspar, Melchior, and Balthazar,— a work of the eleventh century, was ornamented with engraved gems of various kinds. The skeletons of these kings were crowned with diadems of gold and precious stones, with their names delineated in rubies; but these have been replaced by gilt, silver, and pastes.

Probably no shrine of modern Europe has had a wider celebrity for its votive offerings from all classes of devotees,— princes, nobles, priests and peasants, than that of Loretto, in Italy. Among the royal votaries were Queen Henrietta Maria, who presented a golden heart set with diamonds, and Christina,

Queen of Sweden, who offered at this shrine a sacrifice which but few sovereigns have ever been willing to surrender, her crown, sceptre, and jewels. The "Casa Santa," or holy house, once occupied by the Virgin, was borne through the air from Palestine, by angels, and finally settled at Loretto, according to tradition, where it was eventually enclosed in an elegant building ornamented with the works of some of the most celebrated artists of Italy. The statue of the Virgin, sculptured by St. Luke, wears a triple crown of gold, diamonds, and pearls, and a costly robe covered with radiant jewels. The niche occupied by the "Holy Mother" is richly adorned with precious stones; at her left hand stands an angel of silver, and on her right a golden cherub emblazoned with glittering jewels, while gold and silver lamps are kept continually burning before the venerated image. In the treasury of the Casa Santa, is a monstrance, in the form of rays, containing six thousand five hundred and eighty gems.

The most famous English shrine was that of Thomas à Becket, at Canterbury, which, during the zenith of its glory, was visited by one hundred thousand pilgrims in a single year. "It glittered," writes a chronicler of the times, "with the rarest and most precious gems, of extraordinary size"; but nothing now remains of this once venerated object. The shrines of St. Cuthbert at Durham, described as a "blaze of gems," and St. Ethelreda at Ely, were famous in their day; the latter, with the Canterbury and Westminster shrines, contained an immense treasure in gold and precious stones, "enough," says a contemporary writer, "to ransom great kings from captivity." Henry III. presented to the shrine of Edward the Confessor, in Westminster Abbey, munificent offerings in the form of rubies, sapphires, emeralds, pearls, and other gems. This sacred receptacle was a very elaborate and imposing structure of its kind, consisting of two stories, and is an illustration of many others of a similar character. The lower apartment was ornamented with the more common decorative stones, as marble, alabaster, serpentine, and porphyry, while the upper, which enclosed the effigy of the saint, was embellished with the costliest materials,—gold and precious stones; the offerings were usually arranged about the basement of the shrine.

CHAPTER VIII.

PRECIOUS STONES IN LITERATURE.—THEIR MYSTICAL PROPERTIES.

Objects as beautiful and as valuable as gems would, naturally, claim the attention of writers from the earliest times, either as rhetorical figures or as themes for scientific and literary investigation and description. The names of many of these writers have come down to us from antiquity, either as historical or traditional characters, but of whose writings nothing now remains. Pliny cites thirty-six ancient writers on precious stones, yet nothing of all their productions on this subject before his time, is extant except the works of Theophrastus, B. C. 300. The so-called Orpheus, whose "Lithika" has been ascribed by some critics to an Asiatic Greek of the fourth century, was written, thinks Mr. King, by the author of the "Argonautica," Apollonius Rhodius, B. C. 222–181, judging from the style and close resemblance of the two poems.

In the list of early writers on precious stones or those who have referred to their uses, we find the names of Herodotus, Democritus, Theophrastus, Pliny, Zoroaster, Solinus, and Quintus Curtius, besides many of the poets and others of less note. Perhaps of all the ancient writers, none have used them more frequently or more effectively as figures of rhetoric, than those of the sacred scriptures, more fully illustrated in the chapter on "Sacred Uses of Precious Stones." Secular writers make frequent use of them for embellishment; in the "Metamorphoses," Ovid thus describes the palace of the sun: —

"The princely palace of the sun stood gorgeous to behold
On stately pillars builded high of yellow burnished gold,
Beset with sparkling *carbuncles*, that like to fire do shine,
The roof was framed curiously of yourice pure and fine."

According to Palingenius, the "City of the Moon" rivalled in splender the "Palace of the Sun." He says: —

"The lofty walls of *diamond* strong
Were raised high and framed
The bulwarks built of *carbuncle*
That all as fire glowed."

The hall of a magical palace described in the "Gesta Romanorum," was decorated with pearls, diamonds, rubies, and other gems, "glistening like coals of fire."

Marco Polo says precious stones were abundant in Ceylon, a statement corroborated by modern travellers; but for his description of a ruby belonging to the royal treasury, which was a palm in length and of the size of a man's arm, he must have drawn largely upon his imagination. This ruby was undoubtedly a remarkable gem, since it has been mentioned by other travellers, some of whom thought it was a hyacinth, and not a ruby, of the size and form of a pine cone, and placed on the summit of a pagoda so as to be seen at a distance; but the most marvellous precious stone of the old chroniclers was a carbuncle belonging to the King of Pegu, with a brilliancy so penetrating that it rendered the bodies of bystanders transparent.

Sir John Mandeville of the thirteenth century, who saw more wonderful things than most travellers, describes the palace of the Great Khan of Tartary as of wrought gold and precious stones — with hangings of silk, gold, and pearls. The steps to the throne were gold enriched with gems; his table was gold, crystal, and amethyst, over which was suspended a vine covered with fruit of crystal, beryl, topaz, garnet, almandine, emerald, peridot, and onyx; cups, drinking-vessels, and other articles, were made of precious stones.

This eminent traveller visited Ceylon, where he saw the magnificence of the emperor's palace; the royal apartment was lighted by night by a pillar of gold set with a ruby and a carbuncle one foot in length. He entertains his readers with an account of that mythical personage, Prester John, whose magnificent palace in the city of Susa surpassed those of all other oriental rulers in richness and splendor. The throne was ascended by alternate steps of onyx, crystal, green jasper, amethyst, sardonyx, and chrysoberyl, bordered with gold and orient pearls. All the columns were of fine gold set with precious stones, the carbuncle serving the purposes of solar light. The bedstead of this luxurious monarch was made of gold embellished with sapphires. This account of Mandeville differs, in some points, from the description given by another writer, though both authors coincide in regard to the marvellous splendor of the court of Prester John. We are told that the royal palace was built of costly wood and adorned with gold and precious stones; at the extremities of the gable were two golden apples and two large carbuncles; the apples to give brilliancy by day, and the carbuncles to illuminate the night. The gates to this magnificent structure were made of sard, and the windows of crystal; the jousting-hall — this oriental despot, it is evident, had adopted the feudal customs of the West — was paved with onyx, and the furniture of the banqueting-hall was of gold and amethyst. The royal treasures were guarded, night and day, by three thousand armed men.

Passing from the marvellous narratives of early travellers to the no less wonderful descriptions of contemporary poets, we find that precious stones held no subordinate rank in works of the imagination. Lydgate, writing in the fifteenth century, represents in his poem "On the Siege of Troy," the walls of that city twenty cubits high, made of marble and alabaster adorned at every angle with a crown of gold set with the richest gems; the windows of the royal palace were wrought with beryl and crystal. A magical tree twelve cubits in height, whose branches of gold and silver overshadowed the plain, produced blossoms of different-colored gems, which were renewed every day. The Trojans anticipated mediæval architecture by many centuries, since Hector was buried near the *high altar* of the principal *church* of Troy. In Dyer's "Golden Fleece" the palace of Priam was paved with crystal garnished with diamonds, sapphires, emeralds, and other precious stones, while the hall was lighted by an enormous carbuncle set with other gems on the gold crown of a gigantic statue of Jupiter, fifteen feet in height.

Chaucer, Hawes, and Shakespeare all refer to the supernatural brilliancy of the carbuncle — an idea borrowed from Arabian romances. Hawes pictures a hall of jasper, with crystal windows, and roof overhung with a gold vine bearing ruby grapes, which had its similitude some centuries later, at the palace of the Mogul emperors. Spenser plants a golden vine in Mammon's subterranean isle, which yielded hyacinths, emeralds, and rubies. Ben Jonson's "Alchymist" presents us with agate dishes studded with emeralds, sapphires, hyacinths, and rubies; and spoons of amber, ornamented with diamonds and carbuncles. Marbodus, or Marbœuf, Bishop of Rennes, of the eleventh century, in a poem called "Lapidarium," the earliest didactic poem, it is said, since classic times, has given a history of precious stones and their mystic powers as they were accepted in his time; this poem became the "text-book on mineralogy for five centuries." The author derived his theories chiefly from his predecessors, especially Pliny, the Pseudo-Orpheus, and Solinus. The "Lapidarium" treats of the supernatural properties of stones, their color, and some other physical characteristics. Some extracts from this long poem will illustrate the prevailing theories of that period respecting the nature and powers of precious stones, which the discoveries of modern science have proved to be not only erroneous, but absurd.

The virtues and natural qualities of some of them are thus expressed: —

The Diamond.	Hardness invincible which naught can tame; Untouched by steel, unconquered by the flame.
Agate.	Now regal shapes, now gods its face adorn; Such the famed Agate by King Pyrrhus worn;
Alectoria.	It gifts the pleader with persuasive art, To move the court and touch the hearer's heart.
Jasper.	Of seventeen species can the Jasper boast; Of differing colors, in itself a host.
Sapphire.	Fit only for the hands of kings to wear: With purest azure shines the Sapphire rare.
Chalcedony.	Unlike the jasper, of this precious stone Three hues alone are unto merchants known.
Emerald.	Of all green things which bounteous earth supplies, Nothing in greenness with the Emerald vies.
Sardonyx.	The Sard and Onyx in one name unite, And from their union spring three colors bright.
Onyx.	The name of Onyx, as grammarians teach, Comes from the usage of the Grecian speech.
Sard.	Cheapest of gems, it may no share of fame For any virtue save its beauty claim.
Chrysolite.	The golden Chrysolite a fiery blaze Mixed with the hues of ocean's green displays.
Beryl.	Cut with six facets shines the Beryl bright, Else a pale dulness clouds its native light.

Topaz. From seas remote the yellow Topaz came;
Found in the island of the self-same name.
Hyacinth. Midst other treasures to adorn the ring
This gem from Afric's burning sands they bring.
Chrysoprase. As leaves of leek in mingled shadows blent,
Or purple dark with golden stars besprent.
Amethyst. The gem, if rarer, were a precious prize;
But now, too common, it neglected lies.
Jet. Lycia her Jet in medicine commends;
But chiefest that which distant Briton sends.
Black, light, and polished, to itself it draws,
If warmed by friction, near adjacent straws.
Lodestone. The Lodestone peace to wrangling couples grants,
And mutual love in wedded hearts implants.
Coral. Wondrous its power, so Zoroaster sings.
And to the wearer sure protection brings.
Carnelian. Fate has with virtues great its nature graced;
Tied round the neck or on the finger placed.
Carbuncle. Like to the burning coal whence comes its name;
Among the Greeks as Anthrax known to fame.
Lyncurium. Surpassing amber in its golden hue,
It straws attracts, if Theophrast says true.
Aetites. This stone, they say, is found with scarlet dyed;
Hid on the margin of old ocean's tide.
Thunder-stone. From clashing clouds the wondrous gem is thrown —
Hence styled in Grecian tongue the Thunder-stone.
Heliotrope. The Heliotrope, a gem that turns the sun;
From its strange power the name has justly won.
Hematite. Of red and rusty hue, in Afric found,
Or in Arabia, or in Lybian ground.
Hexacontalite. True to its name, the Hexacontalite
In one small orb doth sixty gems unite.
Prase. No virtue has it, but it brightly gleams
With emerald green, and well the gold beseems.
Rock-Crystal. Crystal is ice through countless ages grown
(So teach the wise) to hard transparent stone.
Iris. Its form six-sided, full of heaven's own light,
Has justly gained the name of rainbow bright.
Pearl. Prized as an ornament, its whiteness gleams,
And well the robe, and well the gold beseems.
Malachite. Opaque in hue, with th' emerald's vivid green,
It charms the sight; first in Arabia seen.
Chrysoprase. By night a shining fire; it lifeless lies
Like golden ore when day illumes the skies.

The number of writers on precious stones from the thirteenth to the nineteeth century is very large, but little original knowledge was added to what was previously known, from the time of Pliny, in the first century, to the Arabian writers between nine and ten centuries later. Ibn-Sina, better known as Avicenna, a celebrated Arab philosopher and physician, 980–1037 A. D., who is said to have anticipated the discoveries of modern science, wrote on the subject of precious stones; but Mohammed Ben Mansur, of the twelfth century, was the first after Pliny to compose a scientific treatise on gems, which he dedicated to the Shah of Persia. His knowledge, thinks Mr. King, was marvellous considering the age in which he lived; he was actually in advance of all other writers of his time, and equal to Haüy, Mohs, and others in their supposed discoveries several centuries later.

Michael Constantius Psellus, a writer of the eleventh century, composed a work on the "Virtues of Stones," in which he promulgates a knowledge of their medicinal uses; Jerome Cardan, 1501–1576 A. D., divided all gems into three classes: first, brilliant and transparent; second, opaque; third, compound. He advanced the theory that diamonds and opals originated from gold, sapphires and rubies from silver, and carbuncles, amethysts, and garnets from iron. A work on precious stones, "De Lapidibus et Gemmis," composed by De Boot, or Boethius, is regarded as an example of a remarkable combination of "varied learning and absurd credulity."

Mystical Powers. — Frequent allusions to the popular belief that precious stones are endowed with supernatural powers, and that they exercise some mysterious influence on the fortunes and destiny of individuals and communities, met with in ancient, mediæval, and modern literature, renders some acquaintance with these notions necessary to an intelligent understanding of such references. Endowed as they have been, by the greatest philosophers of antiquity, with life and mysterious qualities, it is natural they should have been invested with godlike attributes, and regarded as a kind of divinity by the superstitious, who are to be found in every age. A belief in the efficacy of precious stones for talismans has been cherished, to a greater or less extent, by individuals among all nations, beginning with the oriental, and spreading westward over the continent of Europe, so that in the Middle Ages it became a universal doctrine, traces of which still linger in certain communities. This popular idea is developed in the "Speculum Lapideum" of Camillo Leonardo, of the sixteenth century.

The magical powers of the so-called diamond ball of Dr. Dee, a contemporary astrologer and mathematician, were generally accredited, and served to strengthen the common belief. This ball, now preserved in the British Museum, is made of rock-crystal and may have been similar to the globe, says King, which constituted the Rhombus or Turbo used by witches in their incantations, and referred to by Horace in the line: —

"Reverse the magic wheel and break the spell."

The cuneiform inscriptions of Assyria refer to seven black stones which personified the same number of planets, objects of adoration in the principal temples. This worship was widely spread in Syria and Arabia, where remains of this superstition still exist. Probably the famous "black stone" of the Kaaba, a sacred shrine in the great mosque at Mecca, is a relic of this practice, since it was an object of popular reverence long before the time of Mahomet, who made a skilful use of this fanaticism to establish and perpetuate his own system, by placing the stone in the northeast corner of the shrine, to be kissed by pilgrims. It is described as a dark basalt or lava, perhaps an aërolite, of an oval shape, four feet in the largest diameter and two in the shortest — some say, seven inches.

Traditions about the wonderful stone of the Kaaba, called the Kiblah or Keblah, assume many different versions. It was one of the precious stones of Paradise, which fell to earth on the advent of Adam, but was lost and subsequently restored by the angel Gabriel. The more popular opinion maintained that it was originally the guardian angel appointed to watch over Adam in Paradise, but changed into a stone and expelled with him for not having been more vigilant. At the resurrection, this stone will assume its angelic form and appear as a witness before God in favor of all faithful Moslems who made a pilgrimage to Mecca. When first placed in the Kaaba, it was a jacinth of "dazzling whiteness," but it became gradually blackened by the contact of polluted lips. The southeast corner of the second shrine is occupied by another venerated stone, which pilgrims are allowed to touch but not kiss.

The serpent has been invested with mysterious powers from time immemorial, which may have been partly in consequence of the gems supposed to be concealed in his head. Some of these subtle creatures have possessed eyes of jacinth, others have been decorated with rings or collars of emeralds, while Milton's serpent had eyes of carbuncle. The famous "Draconius," derived from the head of the dragon, was a black stone which possessed the attributes of absorbing poison and of rendering its owner invincible. The toad "wears a precious jewel in his head"; hence we have a toadstone, an antidote for poison, as well as an indispensable agent in the performance of certain superstitious rites.

Sovereigns and mighty conquerors have not been exempt from the weaknesses of common humanity, we must believe, when we read of their care to possess the Bezoa or Beza stone, obtained from a wild animal of Arabia, and used as a charm against plague and poison. Tavernier alludes to this stone and ascribes its origin to goats and apes. He tells us how to detect the genuine from the false Bezoa, as so important a stone would have its counterfeits. There are two infallible tests: one is to place it in the mouth, and if it is genuine, it will give a leap and fix itself on the palate; the other consists in placing the stone in a glass of water, and if a true Bezoa, the water will boil.

Four Beza stones are enumerated among the treasures of the Emperor Charles V. after his death, and one great Beza stone, set in gold, which had belonged to Queen Elizabeth, was counted among the jewels of James I. With all its wonderful powers, it could not save the Constable of France from a tragical end. Condemned to die, just before his execution, he removed from his neck a Bezoa which he had long worn as a charm, to be given to his son as his dying legacy. It was sometimes prescribed by the medical faculty as a remedy for disease, as in the case of Lorenzo de Medici, but without any efficacy to save the life of the distinguished invalid.

The Bezoa stone, sometimes called the stone of "Jachen," has been represented as of large size and great beauty, of a bluish white color crossed by white veins, and so hard that it could be worked only with diamond powder. It was sometimes cut into cups and other vessels, which were embellished with carved figures and gold ornaments.

Precious stones have not been wanting for nearly all the ills of life, and have been obtained, according to tradition, from the vulture, eagle, swallow, raven, tortoise, hyena, stork, and even the cat. A Peruvian animal, called the "Carbuncolo," of the size of a fox, is invisible in the daytime, yet emits a brilliant light in the dark, from a precious jewel which he carries in his forehead. The Greeks referred all petrifactions to a certain kind of stone which had the power to assimilate every substance placed in contact with it. The flesh-consuming Lapis Assius, used for sarcophagi, would absorb a deceased body in forty days; but Boethius, many centuries later, surpasses the wonders of Theophrastus by his marvellous stone which consumed the flesh of living persons. The eagle-stone, it was thought by some of the ancient philosophers, propagated its species after the manner of animated nature, an idea originating, probably, in the existence of geodes, sometimes called bastard eagle-stones.

The famous lynx-stone has given rise to many conjectures, some writers considering it the belemnite, others amber, hyacinth, or tourmaline. It was valued by the ancients for its medicinal properties, though Pliny discredits not only its healing powers, but also its existence.

The most remarkable stone recorded in history was in the shape of a helmet found in the River Eurotas. This intelligent species responded to the call of the Spartan trumpet to arms, by leaping upon the bank of the stream; but at the sound of an Athenian signal it had the cowardly instinct of jumping back into the water — not a very commendable habit for a Spartan.

Some precious stones were the harbingers of evil: as the onyx, when worn alone, exposed one to danger from malignant spirits; but this objectionable quality was neutralized by combining the stone with sard. The opal has been considered an unlucky gem, an opinion which became very prevalent after the publication of "Anne of Geierstein." The mystical powers of the beryl are not unfrequently alluded to in literature; when it was "charged" in set forms, it received the attribute of revealing secrets, past and future; consequently it was employed in certain rites practised in witchcraft, and might have been used for that purpose by Michael Scot, the famous necromancer of the thirteenth century. Engraved gems representing certain characters were powerful in expelling diseases and exorcising evil spirits; the virtues of the agate for such beneficent purposes were very generally recognized throughout Christendom during the Middle Ages. It is said that a sapphire was kept in the old church of St. Paul, London, for the express object of curing disorders of the eyes.

Most of the legends and traditions about the supernatural powers of gems introduced into Greece and Rome were embodied in beautiful works of art represented on engraved gems, by which the efficiency of these qualities was greatly augmented. A diamond bearing a soldier's head insured victory; a ruby with the figure of an orator was a guaranty for riches and honor; a sapphire representing a musician advanced its owner to a position of great dignity; while a sard or an amethyst engraved with the figure of a warrior strengthened the memory.

It was a common belief that a serpent was made instantly blind by looking at an emerald. Moore alludes to this in "Lalla Rookh" in the couplet:—

> "Blinded like serpent when they gaze
> Upon the emerald's virgin blaze."

The Shah of Persia owns a diamond, so it is reported, which renders him invincible, and another which forces a confession from conspirators,—two important aids for sovereigns of the present day. King Solomon possessed a ruby which gave him power over demons and genii, and revealed to him whatever he desired to know in heaven and upon earth. This learned naturalist says, "Divers are the virtues of stones," which leads us to infer that he accepted the popular theories on this topic. It was the prevaling opinion in ancient times that precious stones were endowed with organic life, that they breathed, and had the power to increase or diminish their size at will. Pythagoras endowed them with souls, Theophrastus with sex, Dioscorides with marvellous powers; Plato believed they were produced by fermentation; and Cardan thought they were subject to illness, old age, and death.

The diamond was one of the most marvellous of all the gems, being propagated, according to Sir John Mandeville, in a manner similar to organic beings; that it was important in defensive armor we are apprised in "Paradise Lost." Chaucer, in his "Romance of the Rose," alludes to the supernatural powers of gems. The San Graal, celebrated in poetry and romance by the writers of the Middle Ages, and the object of religious veneration for many centuries, was a cup made of a single stone, thought to be an emerald, detached from the crown of Satan when he fell from heaven, and was used at the celebration of the Last Supper, and, subsequently, to receive the blood of Christ when expiring upon the cross.

Both Newton and Boyle, two of the most eminent English philosophers, are said to have given some credit to the popular belief in the medicinal qualities of precious stones. If this was true, it was probably in the sense that some other mineral substances are used for curative purposes, and not because they were endowed with any marvellous properties.

In its early history, science was closely allied to superstition, when all the laws and forces of nature were invested with mysterious powers, and no material substance, perhaps, was endowed with so many of these as precious stones. This accounts, in part, for their influence over the imagination during the mediæval period of intellectual slumber, but fortunately modern science has dispelled these fancies, and at the same time has revealed their true character, which presents them in a light no less interesting and remarkable.

CHAPTER IX.

ENGRAVING ON PRECIOUS STONES.

Engraving was well understood by the ancients, and was, probably, one of the earliest of the fine arts, having been practised by the Babylonians, Assyrians, Egyptians, Phœnicians, Hebrews, Greeks, Romans, Peruvians, and Mexicans, and, possibly, by other nations; but this, like all the other fine arts, was lost during the turbulent period of mediæval times, and with them, was recovered at the general revival of learning. None of the ancient arts have been so important to the historian and the archæologist as engraving, on account of its permanence, arising from the indestructible nature of the materials employed. Architecture, sculpture, and painting, are more or less perishable; hence, they have left only partial records of the past, or have been utterly destroyed, leaving no traces of their existence save in the annals of the historian; while gems, on the contrary, owing to their physical qualities and diminutive size, have suffered little or nothing from the ravages of war or political revolutions which could affect the inscriptions they bear. They have, in many instances, been the conservers of the more perishable creations of the sculptor, the painter, the architect, and the poet. They illustrate the myths, legends, historical events, manners, and customs of the early races, and represent the portraits of distinguished persons, the costumes of the different nations, their implements of warfare and domestic use, and their religious rites and ceremonies, all of which could have been preserved so effectually in no other way. There seems to be quite a difference of opinion as to the origin of gem-engraving, some writers having ascribed it to the Ethiopians; some, with more probability, to the Egyptians; while others are divided between the Chaldæans and the Assyrians. Cesnola says the latter afford, beyond all question, the earliest examples of the true process of engraving on hard stones, the Egyptian intagli being merely incised with the graver in much less obdurate materials. On the other hand, M. Perrot believes Chaldæan engraving must have been among the oldest of the kind, if not the earliest; while that of the Assyrians, like their sculpture and architecture, was imported from Babylonia.

The wheel for cutting came into use in Chaldæa about the eighth century, B. C., though engraving on precious stones was understood ages before. By this art, much of their history has been transmitted to posterity, as well as their religious beliefs, represented by the figures of their divinities and sacred emblems carved in cylinders, cones, scarabei, rings, tablets, and other objects. The materials first employed for this purpose were wood, bone, shell, marble, and steatite; later, the harder substances, such as serpentine, porphyry, basalt, syenite, hematite, bronze, and, finally, the same class of precious stones that were subsequently used for engraving among the Greeks and Romans. A fine cylinder in the New York Museum of Art represents Izdubar and Hea-bani, the Hercules and Theseus of Chaldæan mythology, engaged in a hand-to-hand contest with a wild bull and a lion. It was cut on marble, or porphyry, and dates some fifteen centuries before our era.

The production of Babylonian cylinders constituted a national industry, carried on for many centuries; while the cities of Ur, Erech, and Arade, became famous schools of engraving. The Chaldæans, at a later period, made use of seals, cut in hard stones, in the form of cones, pyramids, and spheroids, which were more easily handled than cylinders. The cone was engraved in intaglio, on the base, with subjects less varied than those upon cylinders; but, like them, they had no types in nature, and were such as were represented in the Chaldæan cosmogony. About four hundred of these cones are in the British Museum, and as many more in Paris, nearly all of them cut in carnelian or chalcedony of a fine blue tint.

There is no question but that the Egyptians early acquired the art of engraving, as the Israelites must have obtained their knowledge of it from them during their long residence in Egypt. There are traditions among the rabbies that Chael, one of the Hebrews, while journeying in the Wilderness, engraved precious stones with astronomical signs, and described their history and magical powers, and that Moses engraved the stones of the breastplate with the blood of the worm called "Samir,"* which some writers have interpreted to mean the adamas, while others maintain that the Hebrew word for diamond is derived from a different root, signifying "to smite." The earliest historical engraved gems are generally believed to have been those in the first breastplate of the Jewish high-priest, though Egyptian priests were accustomed to wear engraved tablets when officiating at their religious rites, long before the Jewish ritual was introduced among the Hebrews.

It is supposed the ante-Homeric Greeks were unacquainted with the art of gem-engraving, judging from the silence of Homer, who makes no mention of engraved gems; though one was placed on the finger of Ulysses, by the painter Polygnotus,

* The legend about the blood of the worm Samir, says a modern writer, originated, undoubtedly, from the word "Smir," a material used by ancient engravers.

some centuries later. The Phœnicians, who were the Britons of antiquity, very likely diffused a knowledge of this art among the Asiatic and insular Greeks, for, as early as Homer's time, this commercial and enterprising people traded in jewelry with the islands of the Ægean. However, it was not long after before the art was introduced into Greece proper, where the signet ring became so popular, and its use was carried to such excesses that Solon deemed it essential to the prosperity of the nation to check this extravagance by enacting laws regulating the business of engraving. After the Macedonian conquest, it became very flourishing in Asia, — a region where before it was comparatively but little practised. About the time of Augustus, this art reached its highest excellence, especially in portraits ; and in the reign of Hadrian, it began to decline at Rome, but found, says King, an asylum in the Persian Empire, where it flourished from the third to the seventh centuries, when it suddenly came to an end by the Mohammedan conquest. The religion of the conquerors permitted only cipher inscriptions upon signet stones, which, with their graceful Arabic curves, were very beautiful, and were highly valued throughout the East.

At the Renaissance, the art was revived, and the antique-engraved gems hoarded by amateurs contributed in no small degree to the general revival; while Cinque-cento engravers appeared, whose numerous productions are seen in modern collections. The one at Naples, formerly owned by Alexander Farnese, comprises a magnificent casket of silver-gilt with plaques of crystal engraved with subjects from the history of Alexander the Great ; another example of this school, consisting of a casket of rock-crystal engraved with scenes from the Passion of our Lord, is found in the Florence gallery. The art flourished in Germany under the patronage of Rudolph II., when Nürnberg and Strasburg became the earliest centres of this business.

The "rejuvenated art" was destined to suffer another eclipse during the civil commotions which convulsed Europe, but, though checked in its progress, it was not uprooted, and in the eighteenth century it sprung up with a new vigor, which challenged defeat. The engravings of this period were, to a great extent, copies of the antique; therefore numerous forgeries came into circulation, greatly to the discredit of the art and artists. It has been estimated that for every celebrated antique-engraved gem, a dozen copies exist, so well counterfeited that amateurs and collectors are often deceived; sometimes they even surpass the models in excellence of workmanship.

Engravers. — It is the opinion of connoisseurs that the number of ancient engravers on gems must have been very large, judging from the great number of their productions, though Pliny names only a few, of whom Pyrgoteles, Apollonius, Cronius, and Dioscorides were the best known; to these may be added Aulus, Cneius, Hyllus, and Solon, whose works are found in modern collections. The most admirable engraving of Dioscorides, who is, perhaps, the most skilful and celebrated of the ancient school of art, is the head of Io, in the Piombino cabinet. The signet engraved with the portrait of Augustus, used by the earlier Roman emperors, was a work of this artist.

Many celebrated names are found in the list of modern engravers, whose works have excited the admiration of connoisseurs and collectors. Prominent in this catalogue stands that of Il Vicentino (Valerio dei Belli), whose coffer of rock-crystal, representing scenes in the life of Christ, was purchased by Pope Clement VII., for two thousand gold scudi, equal to more than thirty thousand dollars, and presented to Francis I.,

King of France, on the occasion of the marriage of the Dauphin to Catherine de Medici.

Trezzo, of Milan, acquired an extensive fame for his portraits on gems, his most remarkable effort being the Tabernacle of the Escurial at Madrid, made of different precious stones found in Spain, and upon which he bestowed the labor of seven years. It has been said that he engraved the diamond, but the statement lacks confirmation in the opinion of some writers. Coldorè, who lived in the reigns of Louis XIII. and of Henry IV., enjoyed the reputation of being the first engraver of the seventeenth century. His portraits of Henry, both in cameo and intaglio, display great artistic merits.

Sirletti, an Italian of the eighteenth century, surpasses all modern artists in delicacy of finish, and came nearest, probably, to the ancient Greeks in the artistic merits of his works. The Costanzi were the most distinguished Roman engravers of the present century; Giovanni, the elder of that name, engraved the head of the Emperor Nero on diamond, while Carlo, the younger, produced several masterpieces, — the head of Antinoüs and a Leda for the King of Portugal, on diamond, and the portrait of Maria Theresa, on sapphire; but his most celebrated work was a table emerald, two inches in diameter, engraved with the head of the reigning pope on one side, and of Peter and Paul on the other, intended for a brooch for the pontiff.

Rega, of Naples, who lived in the latter part of the eighteenth century, came nearer the antique style, it is thought, than any other modern engraver. Hercules at repose, and the head of a Bacchante, represent some of his most celebrated works.

Pichler, also a Neapolitan, and one of the first of modern artists, has produced works of the highest merit; his intagli have frequently been sold for antiques.

Pistrucci, a Roman by birth, but for many years a resident of London, was one of the most fortunate of modern engravers in a pecuniary sense; his works were always eagerly sought, and commanded high prices. His Flora, bought for an antique, was considered the choicest gem in the cabinet of the purchaser, Mr. Payne Knight; his cameo of a Greek warrior on horseback, with slight changes, was adopted for the reverse of English crown pieces and sovereigns at the re-coinage in 1816. The improved copy of the design for a subsequent coinage, during the reign of George IV., is regarded the finest on any modern currency; his heads on the obverse are less successful.

Of English engravers in the last century, Smart was the most notable for the celerity with which he wrought: he is said to have engraved several stones in a single day, and by no means in a careless manner. Seaton was characterized by extreme finish, but lacked spirit; his most famous productions are the portraits of Pope, Inigo Jones, and Newton. Marchant left some fine works in the Greek style, but their finish is too minute for effect, the consequence of using the microscope in the process of engraving. Mr. Streeter says that Italy, France, and England afford the best engravings of modern times, and the imitations of antiques are so perfect that it is with difficulty they can be distinguished from the genuine, even by experts.

The practice for engravers to affix their names to their works did not come into use until the time of the Emperor Augustus. A Diana on sard, in the archaic style, belonging to the Stosch collection, is supposed to be the oldest gem known bearing the artist's name — Heias.

Subjects for Engraving. — The themes for the engraver's art are limited in variety, consequently they are frequently repeated. A large proportion represent classical scenes; the remainder comprise portraits, representations of animals, the eagle being the favorite, certain implements denoting occupations or social customs, masks, chimeras, and other whimsical fancies. The heathen divinities afforded an attractive subject for art with the early engravers, but later, the portraits of royal persons were substituted for those of the gods, a practice adopted about the period of Alexander the Great. Engraving portraits on gems was not in vogue before the Macedonian princes, who set the example by placing their own heads upon coins instead of that of the tutelar divinity, as had been the custom. The Greeks represented their gods in human form, which was not the case with most other pagan nations, whose personifications embraced some of the most grotesque figures the fertile imagination of the oriental mind could invent. The Greeks never used the beetle, so commonly represented by other nations, but either single divinities and heroes or groups illustrating some scene in Homer or the tragic poets.

The ancient Romans generally selected their subjects from the scenes of ordinary life — war, hunting, agriculture, or some religious ceremony, but never from the poets. Portraits of the emperors appear in the earlier imperial times, but rarely at a later date, and then generally in the character of Mercury with the caduceus. The signs of the zodiac, designating the horoscope of the possessor, were favorite subjects for engraved gems; Capricorn, often seen with the portrait of Augustus, was believed to designate high dignity and power. At a later period, this kind of engraving was worn as amulets for protection against disease and accident. A favorite subject with the Egyptians and Etruscans was the beetle, and the gems engraved with this insect were called scarabei, or beetle-stones; they are considered among the oldest monuments of the glyptic art in existence. These engravings were at first executed on soft stones, as steatite and limestone, but as the art advanced, harder materials, such as basalt, agate, lapis-lazuli, and others, were employed. Egyptian scarabs, though more ancient than Etruscan, were inferior in workmanship until the time of the Ptolemies, when, it is said, the Greek style was engrafted upon the old stock, producing a new scion, called Greco-Egyptian, which has afforded some fine specimens; one of these is seen in the British Museum, and another in Berlin.

It is supposed the Etruscans adopted the arts and religious system of the Egyptians — though some antiquaries believe their scarabs exhibit traces of Asiatic origin — hence we find among the remains of this ancient people, these emblems, cut in a great variety of material, from emerald to amber, and even in pastes. The scarabs of Egypt were of all dimensions, from colossal to very minute, while those of Etruria were nearly all of the same size. The divinities of both nations are represented with wings, and are less graceful and more exaggerated in action than those of the Greeks. Etruscan antique ornaments composed of scarabei are sometimes discovered of considerable value; a necklace of this description, found in 1852, in Tuscany, was sold for eight hundred dollars.

The Greco-Italian engravers selected for their subjects, at first, some of the lower animals, as the ox, stag, and lion, afterwards human figures in full length, succeeded by heroes and demi-gods, and, finally, the superior gods, with the forms of men. The favorite topics for Italian art in every department since its revival have been taken from the Æneid and the Metamorphoses; those of the Cinque-cento period were derived from Roman history and the poems of Ovid. Byzantine art affords some examples of engravings of large size but inferior workmanship and tasteless designs. Certain gems with inscriptions in Hebrew or in Persian claim to be antiques, but this circumstance alone does not constitute positive proof of great age.

Style and Character. — Antique art never offends the moral sense by representing degrading scenes, as is sometimes done by modern artists, greatly to the shame of our vaunted superior civilization; but they were those which "custom and reverence sanctioned." The earliest Greek engravings are in low relief, executed with the diamond point, it is believed, and with an Etruscan border, which has led some connoisseurs to assign them to the Etruscan school, — a mistake indicated by the material selected for the purpose. With the Etruscans and early Italians, carnelian was the favorite, while the Greeks preferred a yellow sard resembling topaz, and sometimes amethyst and jacinth. The latter people gave great attention to details, representing hair by innumerable fine lines, all distinct from one another, and never crossing, while the Romans, who aimed at effect, expressed it by broad masses like paintings; short, curly hair was delineated by holes drilled close together, similar to that seen in some archaic marble statues; their portraits exhibit a stiffness not observed in Greek workmanship. The details of the early Roman engravings were executed by the diamond point, but those of a later period were done entirely by the wheel, which, it is thought, came into use in Rome about the time of the Emperor Domitian, A. D. 81–96. The most highly finished intagli display a brilliancy which has suggested the idea that they were cut and polished by the same operation; in modern times, this result is achieved by a complicated process, explained in another chapter.

The Cinque-cento engravers copied the Roman style, with exaggerations. Their earliest productions, executed under the patronage of Lorenzo de Medici, are distinguished by their extreme stiffness and mediæval character, quite in contrast with the flowing style of a later period.

The numerous forgeries extant had an unfavorable influence, writes King, upon the public fancy for engraved gems, and as a consequence the business passed from skilled artists to mere mechanics. After a career of thirty centuries, says this author, the ancient art of engraving upon precious stones may be said to have passed away.*

Intaglio. — This is the form of cutting a gem with the figure depressed, and is opposed to cameo, which represents the design raised. The number of antique intagli still in existence is incalculable, owing to the vast quantities produced during many centuries all over the civilized world, and from the indestructible nature of the material, which neither time nor the elements can affect. They were very numerous in Rome, but few intagli, compared with the countless number of camei, were produced during the Cinque-cento period. Those representing purely Christian subjects, of undoubted antiquity, are, it is said, exceedingly rare, though modern works of the kind are not uncommon. The British Museum contains some of these early Christian gems, including a red jasper set in gold, with an inscription, an emerald engraved with the figure of a fish, and a large sapphire with the monogram of Christ. The Greek intagli were frequently set in finger-rings for signets, the Ariadne in the Pulsky collection affording one of the finest examples of the kind. The intaglio was superseded by the cameo, though in the last century there was a revival of the fashion for intagli, and many were executed equal to the best of ancient workmanship.

* This statement needs some qualification, since modern engraving is practised, to a limited extent, at least, very successfully.

Cameo. — The origin of the name has been referred to different sources, — to the Arabic "camaä" (an amulet), to the Greek "kauma" (heat), and with more probability to chama (a sea-shell used for camei); the earliest adoption of the term for figures in relief was in the beginning of the sixteenth century. Antique Roman camei are nearly all of large size, and not intended to be worn as ornaments, while, on the contrary, the Greek specimens were seldom above the ordinary dimensions. It is claimed by connoisseurs that shell and turquoise antique camei are extremely doubtful; the busts of the Cæsars in shell, contained in the South Kensington Museum, belong to the early Renaissance. This form of engraving is generally cut on opaque or translucent gems, while intagli are more frequently found on transparent stones. Antique camei on sardonyx were usually in three colors, if the layers occurred in regular succession, with the base of a translucent dark chocolate, the middle opaque white, and the upper layer a light brown or red. Sometimes the head of a warrior was cut in red, the helmet in green, and the breastplate in yellow, a rare combination of colors.

Cameo is much later than intaglio, and since the Renaissance the number produced has been vastly greater than the latter, partly, no doubt, because they can be executed with greater facility. The Republican period of the Romans has been called the age, par excellence, for camei, while those of Grecian work are extremely rare. Some of the best antique specimens date from the reign of Hadrian, in the second century of our era, after which the art began to decline. The oldest known cameo is said to be the Ptolemy and Berenice, on sardonyx, in the Odescalchi collection. The Romans of the present day, who make use of the Indian conchs, have carried the art of shell camei to a surprising height of excellence.

Rome, said to export fifty thousand dollars worth of these jewels annually, possesses almost a monopoly of the trade, though Paris has produced some fine specimens, notably a sardonyx, which was sold for seven thousand francs. One of the most celebrated camei of modern times is an onyx three inches in diameter, representing Shah Jehan in the act of killing a tiger, during a hunting excursion.

This kind of work has been very freely used for decoration in various ways, both in ancient and modern times. The shrine of the Three Kings of Cologne is ornamented with camei, and some of the vases in use among the Romans were entirely covered with them ; such is the famous "Agate of St. Denis." Francis I., King of France, was the owner of an agate vase three inches in height and two in width, cut in three different colors, and engraved with the figures of Apollo and Diana, Cupid and Psyche, and Victory seated on a car drawn by butterflies, mounted in gold and enriched by precious stones. During the reign of Louis XV., this remarkable vase was despoiled of its ornaments and sold for a trifle. The Tazza or Farnese Cup, at Naples, eight inches in diameter, is carved in the interior with reliefs, supposed to represent the "Prosperity of Egypt," and on the outside with the head of Medusa. The Apotheosis of Augustus, on a tri-colored sardonyx, thirteen inches by eleven, is one of the most celebrated and, perhaps, the most superb monument of the glyptic art in existence. It comprises twenty or more figures, and is in two parts. It was brought from the East by the Crusaders, in the reign of St. Louis, and dedicated by Charles V., King of France, to the chapel in his palace ; it is sometimes called the "Onyx of Sainte Chapelle," and is to be seen at the Bibliothèque Nationale, in which are preserved a large number of camei and intagli. A cameo representing the four horses of a quadriga, in four different colored layers, is a rare and interesting specimen, on account of the number of tints and their skilful use by the artist. The Odescalchi cameo, in the Vatican collection, remarkable for its size and superior workmanship, is engraved with figures once supposed to represent Alexander and Olympia, but according to the opinion of Visconti, Ptolemy Euergetes and Berenice. The largest cameo in this cabinet delineates the triumph of Bacchus and Ceres, on a gem sixteen inches by twelve, consisting of five different colored layers. Another of large size and superior workmanship, cut upon a stone of two colors, is in the Vienna Museum.

The Florence collection comprises many celebrated engravings; some of the most conspicuous are an Antoninus Pius, a Cupid on a Lion, Apollo in Repose (the figure being in gold), Iphigenia recognizing Orestes and Pylades, a head of Jupiter, a Bacchante cut in three colors, the head of Augustus as Apollo, head of Vespasian, head of Livia, Wounded Stag, and the Fall of Phaeton. Of sacred subjects, are a figure of Christ, of large size, and a double cameo, in blood-red jasper, depicting the Flight into Egypt on one side, and on the other the Massacre of the Innocents. The group of portraits affords five of the Medici family — Cosmo, Lorenzo, Alexander, Catherine, and Leo X.,— Francis I., Phillip II., and Bianca Capella. Those of the fifteenth century include Savonarola, on carnelian, and Pope Paul III., on sapphire. A few of the most celebrated intagli in this collection are an Apollo, on onyx; Hercules in Olympus, on amethyst; Titans, on amethyst; Pallas and other heads, on sardonyx; and Leander, on sapphire. A large number of the Florence engravings bear the name of Lorenzo de Medici, who established a school for this department of art in the Republic.

A few Indian and Persian camei of antique workmanship are in existence, including, as one of the most notable, the representation of a Sassanian monarch, supposed to be Sapor II., the production of an Asiatic Greek. The portraits of this ruler, whose reign extended through seventy-two years, are very numerous. An intaglio upon emerald, with the figure of Cupid teasing a goose, and an aquamarine engraved with Cupid on a dolphin, both in the British Museum, are regarded by judges as specimens of exquisite workmanship.

Busts and statuettes were frequently carved from solid gems by the Romans. One of this character—a bust of the Emperor Tiberius—occurs in the Florentine collection. Pliny refers to the statue of Queen Arsinoë, four cubits in height, made of topazion, which may have been peridot or agate.

Collections of Engraved Gems. — The best public collections of engraved gems, says Mr. Streeter, are in Berlin, Vienna, Florence, Naples, St. Petersburg, Copenhagen, and London. The British Museum is said to contain specimens of the finest and rarest types of gem-engraving. For special collections, the most worthy of note are those of the Barberini Palace, and of the Duke Odescalchi, at Rome, and the Blacas, in the British Museum, which comprises some of the most valuable in the world; but by far the largest number of these monuments of art are to be found in the cabinets of noble and wealthy amateurs.

One of the most extensive collections of engraved gems in the United States is found in the Metropolitan Museum of Art, New York, referred to in the chapter on "Collections of Precious Stones." These engravings, if not so numerous as those of some other countries, are valuable for the variety they offer, as well as for the excellence of the stones themselves. From this collection, a few specimens are selected, to give some idea of its general character, more especially in the subjects chosen for representation. Assyrian art is typified by the cylinder, the Persian by the cone. One of the cylinders, made of black hematite, is engraved with the figure of the god Belus, having the winged disc above his head, the symbol of the divine presence, and holding in his hand the crux ansata, the emblem of life — the real origin, it is thought, of the English royal orb and cross. Several human figures, animals, and emblems, are introduced, making the whole scene a complex "mixture of Assyrian and Egyptian ideas." The bust of a Sassanian king, on garnet, is represented with the usual pearl of immense size in his ear, understood, from an inscription in Pehleve characters, to be Sapor II.; the bust of a queen, with a strikingly marked type of national character, engraved on lapis-lazuli of very superior quality, is thought to be that of an Indo-Scythian. The Egyptian ideas are represented by Horus, one of the sun-gods, seated on a lotus, the emblem of fertility, with a star and triangle, executed upon green jasper; while the Gnostic doctrines are represented by the Abraxas god, corresponding to Serapis, in various forms. One of the Greek gems — a sard of different shades — bears the head of Saturn, — a rare subject for engraving; and a fine work on red jasper exhibits the figure of Faustina the Elder, as Cybele, one of the earliest known on that material; Jupiter, Juno, Apollo, Minerva, Serapis, and other deities, of Greek, Roman, and Egyptian mythology, are all represented, with different degrees of skill, on a large variety of precious stones.

Kinds of Precious Stones used for Engraving. — Sards of different shades constitute more than half the number of engraved gems; the remainder comprise several species, including the following: —

Ruby. — Some connoisseurs deny the existence of any real

antiques in ruby or carbuncle; but King mentions a few examples which date from a late period of the Roman Empire. The Devonshire parure comprises a Venus on this gem.

Sapphire. — The hyacinthus of the ancients was a favorite with them, and was frequently used as an ornament without cutting; but it was sometimes engraved. The most celebrated of this kind known was the signet of the Emperor Constantius II., of the fourth century, weighing fifty-three carats, now in the Rinuccini collection. The Marlborough contains two of these gems, one being of a deep violet color, engraved with the head of Caracalla, and the other a sky-blue, carved with the head of Medusa. The gem, hitherto supposed to be sapphire, engraved with Berenice, is believed to be iolite.

Emerald. — True emeralds with genuine antique engravings are said to be exceedingly rare; hardly an example can be found earlier than the second century. Mr. Westropp considers an Egyptian emerald in his collection the only genuine antique engraved gem of the kind; but Mr. King names several of this class: the portrait of Hadrian, the head of Sabina, a Gorgon's head in the Devonshire parure, the Solar Lion in the Fould collection, and the head of Jupiter in that of the Duc de Luynes.

Beryl. — This gem was not often engraved by early artists, though there is a Hercules of Roman date in the Blacas collection, a Julia on a large size aquamarine in Paris, and a Taras on a dolphin, an Etruscan work. The *spinel* affords the head of a Gorgon; and the *balas* one of a Bacchante, considered a masterpiece of the best days of Roman art.

Garnet. — The Greeks seldom employed this precious stone for engraving, though it was frequently used for this purpose by the Romans, and later by the Persians for the portraits of the Sassanian kings. Many fine engravings are found on *essonite,* including a Julius Cæsar, by Dioscorides, and the Apollo Citharidus, both in the Blacas collection; the head of a king of Pergamos, in the Florence cabinet; and a Mæcenas, by Apollonius, formerly belonging to the Herz collection. The antiquity of a Sirius, on Indian garnet, belonging to the Marlborough gems, has been questioned. Antique camei and intagli cut in garnet are often called jacinths, a variety of zircon; but it is claimed there are no antique engraved gems of this species, an evidence that it was not known to the ancients.

Lapis-lazuli. — This gem, supposed to be the sapphire of antiquity, was seldom engraved by the Greeks, but frequently by the Assyrians, Egyptians, Persians, and Romans, as well as by the Cinque-cento artists.

Turquoise. — Nearly all the engraved gems of this species called antiques belong to the Renaissance; a few are genuine, including some of the Persian stones belonging to the Sassanian period. The green turquoise was preferred to the blue, though instances of the latter occur in the Marlborough collection. There is a bust of Tiberius on this gem in Florence, and the mask of an Indian Bacchus in the Blacas cabinet.

The *topaz* of the ancients has no genuine engraved specimens; and the *chrysolite* was seldom used for this purpose, though frequently employed by modern artists. An example of an antique is found among the Townshend gems, at South Kensington.

Pearl. — The ancients rarely engraved this beautiful gem. Two examples are mentioned by antiquaries, — one, representing the heads of Sol, Jupiter, and Luna, supposed to date from the later Roman Empire, and another belonging to the thirteenth century.

Rock-Crystal. — Though this mineral was extensively used

for cups, goblets, and vases, as the unusually large number of antiques in this stone proves, yet it was not employed for engraving until the Renaissance, when it came into use for jewelry. Specimens of engraving on crystal, the work of this period, are seen at Naples, Florence, and in other modern collections.

Amethyst, of the quartz species, was an attractive stone for the engraver of nearly all nations and periods, judging from the Egyptian, Etruscan, Greek, and Roman intagli in this variety. A few of the most celebrated examples are: Omphale, in the Marlborough collection; Atalanta, in Berlin; Achilles, in Paris, Pan, in the Blacas; Sapor I., in the Devonshire; and Mithridates, in Florence. The yellow quartz did not please the ancients as a material for the glyptic art, but the engravers of the fifth century, however, adopted it as a favorite. One antique specimen is found in the British Museum.

The *sardonyx*, usually comprising three colors, is well adapted for camei; therefore, some of the largest engravings known are cut in this gem; the colossal specimen in the Vatican Museum surpasses all others of the kind in size.

The *onyx* was frequently engraved, a remarkable specimen being afforded by the Corinthian helmet on jasper-onyx. An imitation of this engraving constituted the chef-d'œuvre of the notorious Poniatowsky collection.

Plasma. — There is an intaglio on this gem, in the British Museum, of great artistic merit. This kind of engraving was executed in the Roman period on a rare, translucent plasma called by Pliny green jasper. There are said to be no antiques on *prase*, but a large part of ancient sigils are cut on jasper.

Red jasper. — Two of the best known engravings on this variety of quartz are the head of Minerva, at Vienna, and that of Vespasian, in the Marlborough collection. The Romans employed green and bright vermilion jaspers for their works, but the Egyptians and the Gnostics preferred the yellow. They frequently selected the *heliotrope* or blood-stone for talismans, but very rarely for engraving. Some good specimens of antiques occur on *nicolo,* including an Apollo, belonging to the Herz collection, and the head of Caracalla, to the Blacas.

Sard. — Early engravers had a decided preference for the sardius, and some of their best works appear on this stone. The Greeks chose the yellow tints, the Romans the red, though a variety called sardine, of a deep red color, was employed both by the later Greeks and the early Romans, and also by the artists of the fifteenth century as well as by modern engravers. A fine specimen of engraving on this gem, representing a Bacchante, occurs in the Blacas collection.

Carnelian. — The most ancient Egyptian and Etruscan intagli are found on this variety of precious stone, but after the conquests of Alexander, it was superseded by the oriental sard. Its color, toughness, and capacity for polish, render it a desirable material for engraving.

Chalcedony. — This variety of quartz has been sought for engraving by artists of all times; it was used by the nations of antiquity for Babylonian cylinders and for Etruscan scarabei, and has been also a favorite with mediæval and modern engravers. When of a yellowish tint, it was called *opaline,* a substance distinct from opal, a species not found among antique gems. It is estimated that a large part of engraved chalcedonies, notwithstanding its frequent use by the ancients, are modern counterfeits. *Obsidian* and *flint* were substances used by the Assyrians and Egyptians in their earliest attempts at engraving.

Some idea of the relative proportion of the different precious stones used for this purpose may be obtained from the following list copied from the inventory of the Mertens-Schaffhausen collection, which comprised 1475 gems:—

Sard and carnelian	604	Nicolo	49	Chrysolite	4
Chalcedony	274	Amethyst	36	Beryl	3
Jasper	161	Lapis-lazuli	32	Turquoise	3
Onyx	109	Jacinth	22	Ruby	2
Plasma	101	Emerald	10	Opal	2
Garnet	54	Crystal	8	Sapphire	1

Mawe says the carnelian, onyx, agate, etc., could not be engraved by any stone except the diamond. This is an error, since they have been cut by the corundum, and, according to some writers, by copper tools tempered to excessive hardness, an art not understood at the present time. It is a question with some antiquaries whether the diamond was known to the ancient Mexicans, yet they engraved on hard stones.

Gnostic gems. — The Gnostics combined both pagan and Christian ideas in their system, which has been very fully represented on engraved gems, and as Alexandria was the nursery of Gnosticism, the larger part of what are called Gnostic gems are of Egyptian origin.

The worship of the Egyptian god Abraxas, and the Persian god, Mithras, — both representing Phœbus, or the sun-god, — was introduced into this system, hence the title Abraxas has been applied to the entire class of Gnostic gems.

Abraxas, thought to be identical with Serapis, combines various attributes and is represented with the head of a cock or a lion, a human body, and legs composed of serpents. He bears the scourge, the royal emblem of Egypt, and a shield, and was lord of all inferior spirits. The letters when employed as Greek numerals make the number three hundred and sixty-five, the number of successive emanations of the great creative principle, or, according to the Persian Mithraic rites, the first emanation of Ormuz. With the Gnostics, Abraxas was the type of Christ as the Creator and Maintainer of the universe.

The Egyptian Mithras-Abraxas, or Serapis, is represented on intagli seated upon a throne with a triple-headed animal by his side, and Isis, the earth, before him. The engraving bears the inscriptions, "There is but one god, and he is Serapis," and, "Immaculate is our Lady Isis." The Gnostic rites bore a resemblance to some of those pertaining to the Christian religion, as baptism and the Eucharist; they were practised at Rome for a long time under the Christian emperors, and were frequently represented on engraved precious stones. The neophyte when initiated, was subjected to twelve degrees of trial or torture during his probation, which lasted forty days. He was scourged for two days, and compelled to lie naked on the snow for a certain number of nights. These tests are represented on the bas-reliefs in the museum at Innsbrück.

The column, triangle, and some other symbols frequently occur in Gnostic engravings; some of these emblems, as the serpent, sword, level, column, and the name of Saint John, whom this ancient sect claimed as their special apostle, have been adopted by the modern society of Freemasons. The angel Michael is represented on Gnostic intagli as a winged youth with a hawk's head, and holding in each hand a mason's level, with his finger on his mouth, betokening secrecy on the initiated.

The early Gnostic intagli sometimes represented a serpent with a lion's head surrounded with seven rays, sometimes Anubis, a god with the head of a jackal, and sometimes Osiris, wearing a crown. These intagli are the only glyptic monuments existing, says Mr. King, of the later periods of the

Roman Empire, and represent art at its lowest ebb. Most of the Gnostic gems were designed for amulets, or to be carried about as credentials. They were placed with deceased bodies in tombs for safety against the power of demons, and are found in great abundance in the ancient cemeteries of the Gauls. At present, they are numerous in France and Italy; some of the finest Gnostic intagli are in the British Museum. It is believed there are traces of Gnosticism among the sects in the valley of the Libanus mountains, in Syria.

Cuphic gems. — These are precious stones engraved with legends in the Cuphic or square Arabic characters arranged to represent a cross or the letter T; they were not used later than the thirteenth century.

Raspe, in his catalogue of engraved gems, including both antique and modern, of which there is any anthentic knowledge, places the whole number at 15,833.

A list comprising the names of gem-engravers and a description of their works, together with the collections in which they are found, has been prepared by Count de Clarac.

CHAPTER X.

THE DIAMOND.

It is generally conceded that the diamond, including all its varieties of color, holds the first rank among precious stones for beauty and intrinsic value, though at the present time it falls below some other gems in price as an article of merchandise; there are none that equal it in hardness, brilliancy, and a remarkable play of colors.

There is no other object in the whole realm of nature which has been so eagerly sought and so reluctantly yielded as the diamond; and for this reason it possesses great tragical and historical interest, having been not unfrequently the cause of wars, the subject of negotiations between nations, and the incentive for the commission of horrible crimes.

The diamond has always been regarded the symbol of rank, power, and wealth; hence, it has been freely used to embellish royal crowns, and other insignia of distinguished birth, as well as for personal ornaments in the circles of fashion. Perhaps in no period of its history has it been so generally employed in jewelry as at the present time.

This incomparable gem has also served the higher purposes of art, science, and literature. Its imperishable nature affords an appropriate material for the engraver's art; and its remarkable physical and chemical properties render it an object of interesting experiments in science. It constitutes one of the most appropriate and expressive metaphors in literature for whatever is transcendent in beauty and excellence in the whole realm of nature. Diamonds and stars, in rhetorical language, have become almost synonymous terms; and when we gaze at the sparkling firmament in a cloudless night, we perceive how appropriate the comparison between the celestial hosts that "bedeck the sky" and this peerless gem.

Antiquity. — There is much uncertainty as to the earliest period in which the diamond was used as an ornamental stone. It is mentioned by the sacred writers, though some commentators believe a different gem is meant by the word or words translated diamond; while others say the Hebrew word Jahalom means the true diamond. The name may have been derived from the Greek diaphanes, transparent, or adamas, untamable, unconquerable. The latter term was used as an epithet by the early Greeks, as in the dramas of Æschylus, who employed "adamantine chains" to bind Prometheus. It was applied by Theophastus to various hard substances, including emery and other forms of the corundum.

Some antiquaries think the adamas of Pliny comprised the true diamond; while others believe this gem was unknown to him. There are arguments favoring both sides, drawn from the descriptions of this naturalist. He says it was hexagonal in crystallization, — a form the corundum assumes, and not the diamond, — and that it resisted a blow on the anvil, — a mistake later writers have also made; that it was the most valuable substance in nature, which but few possess, even among kings; and that it was discovered with gold, — a description which, in the main, applies to the modern diamond. In his time, the adamas was used for engraving; but it has been assumed by certain writers that the diamond was not employed for that purpose until a later period, though the corundum was thus used in the East at a very early date, and was introduced into Europe under the name of adamantine spar.

It is recorded by the Prophet Jeremiah that the sin of Judah is written with a pen of iron and the point of a diamond. Whatever this stone may have been, it was mounted, like the diamond of the present day, in iron tools for engraving. The prophet's pen may not inaptly be regarded as a type of our modern *so-called* diamond-tipped pens. Manlius, in the first century of our era, is supposed to be the earliest writer who definitely mentions the true diamond so as to leave no doubt about its identity.

Constituent.—The diamond is pure crystallized carbon, the "highest development of a physical substance from a simple element." There is only one other mineral of the same nature, and classed in the same group; the ruby, the sapphire, or the emerald, it might be supposed, is alone worthy a place beside this unrivalled gem. Every one has noticed the black, sooty, greasy-like substance called graphite, used for stove-polish; this unattractive substance is the twin brother of the diaphanous, sparkling gem we call the diamond. Yet no two objects could be more unlike in appearance. Or, if one is not quite assured of the vegetable origin of graphite, let him take a piece of charcoal and notice how soft and lustreless it is, how easily it crumbles, leaving its darkened trace upon everything it touches, and then compare it with the brilliant glittering upon his finger, so hard, and pure, and transparent; are they much alike? yet both are composed of the same element—carbon. Lavoisier was the first to establish the real nature of its constituent, by burning a diamond.

Origin.—The origin of the diamond has been a fruitful topic for speculation among scientists, hence many contradictory theories have been advanced and argued with some show of reason; but after all that has been said and written on the subject, we are still left pretty much in the dark. Theories answer a good purpose, since they often lead the way to truth. But this is not all; they illustrate the ingenuity of the human mind in seeking to account for the methods Nature takes for the accomplishment of her secret operations. Some of these theories about the origin of the diamond are very ingenious and interesting, though the amount of truth they embody remains to be proved. It has been suggested that the vapors of carbon during the coal period may have been condensed and crystallized into the diamond; and again that itacolumite, generally regarded as the matrix, was saturated with petroleum, which, collecting in nodules, formed this gem by gradual crystallization. Newton believed it had been a coagulated, unctuous substance, of vegetable origin, and was sustained in his theory by many eminent philosophers, including Sir David Brewster, who believed the diamond was once a mass of gum derived from certain species of wood, and that it subsequently assumed a crystalline form. Dana and others advance the opinion that it may have been produced by the slow decomposition of vegetable material, and even from animal matter. Burton says it is younger than gold, and suggests the possibility that it may still be in the process of formation, with capacity for growth. Specimens of the diamond have been found to enclose particles of gold, an evidence, he thinks, that its formation was more recent than that of this precious metal. The theory that the diamond was formed immediately from carbon by the action of heat is opposed by another, maintaining that it could not have been produced in this way, otherwise, it would have been consumed. But the advocates of this view were not quite on their guard against a surprise, for some quick-witted opponent found by experiment that it will sustain great heat without combustion.

Later opinions incline to the hypothesis that the diamond

originated in some pre-existing form of carbon, which has been explained as the result of the crystallization of that element from a liquid solution. Carbonic acid collected in cavities, it is affirmed, liquefied under great and long-continued pressure, during which it dissolved some of the pre-existing carbon, when the acid escaped and crystallization began. After the pressure abated, the evaporization of the liquid left a mass of carbon, which constituted the diamond. Another school of scientists teach that this gem had its origin in mud volcanoes, which is in direct opposition to its glacial source, a theory which has also its supporters. In the face of all these conflicting views; we must let the subject rest until some new discoveries afford a basis for fresh speculations.

Matrix. — As many clashing opinions have been advanced about the real matrix of the diamond as about its birth; but after the testimony of different explorers has been sifted, it appears as if there were several different rocks which may be regarded as its native home. The Indian gem is said to occur in a sandstone breccia composed of jasper, quartz, chalcedony, and horneblende, cemented by a silicious substance, the conglomerate, passing into loose pudding-stone, forming the diamond beds. Professor Liversidge says granite, itacolumite, jasper, and peridot indicate the presence of the diamond in India and Brazil, while in Australia it is associated with sandstones, shales, conglomerates, and trap-rocks. Agassiz thought the diamond-bearing formation was the glacial drift, but the rock about Diamantino is itacolumite, which is an intimation that it constitutes the true matrix. We are not left the consolation that our researches end here, since the diamond, like the "Wandering Jew," has been driven from this retreat, to resume its nomadic life. A tradition, says Mme. Barrera, prevailed in the East, in early times, that diamonds were obtained from inaccessible valleys by throwing pieces of meat into them for the vultures to seize and bear away, when the gems adhering to the flesh were scattered by these birds of prey, to be gathered by the natives.

Properties. — The diamond differs from all other precious stones in some of its properties, and surpasses them all in others. It holds the highest rank in the scale of hardness; consequently, it can scratch all other minerals and still remain intact from abrasion. Its specific gravity, though not the highest, is remarkably uniform.

The lustre is superior to that of any other gem, which, however, requires artificial methods for its development in the highest degree, though occasionally it has been found with surfaces naturally polished. This lack of brilliancy in its rough state has, undoubtedly, caused it to be overlooked by those seeking for it.

The remarkable lustre of the diamond is said to arise from its reflecting power, by which it throws back all the light which strikes it at an angle exceeding a little more than twenty-four degrees. This gem possesses the property of bending the rays of light out of their regular course and scattering them; to these refractive and dispersive powers is due that beautiful play of colors, sometimes called its "fire," which constitutes one of its chief attractions. There is no other gem comparable to the diamond for this interesting phenomenon, and it is even possessed in a greater or less degree by different varieties of this precious stone, the perfectly colorless specimens exhibiting it in the greatest perfection, and the darkest in the least, or not at all. The kind of light by which it is seen has an influence upon the vividness of the prismatic colors, artificial being more favorable for producing the best effect than natural, and the flame of a wax candle superior to the brilliancy of gas.

The diamond is a non-conductor of electricity, while graphite and charcoal, substances identical in chemical constituents with this gem, are very active conductors. Both in a natural and polished condition, it acquires positive electricity by friction, while, on the contrary, most precious stones are negative in the rough, and positive only when polished. It does not possess double refraction; neither does it polarize light, as do some other gems. But it is the exceptional one that exhibits phosphorescence in a natural state, and then only in the case of certain stones, though this property is said to be generated by steeping in hot water.

During the last century, it was shown that double refraction never occurs in non-crystallized substances, nor in crystals of the cubic system, to which the diamond is allied. Haüy confirmed this opinion, and maintained the converse, that all crystals not of the cubic system were double-refracting.

No solvents, not even acids, have the slightest influence in decomposing the diamond—a fact which enhances its value as an ornamental stone. Until the middle of the seventeenth century, it was believed to be incapable of injury from heat, but Sir Isaac Newton believed it was combustible, before it was submitted to the operation of burning, on account of its refracting power. The relative density of quartz and diamond are as three to four, while their refractive powers are as three to eight.

The experiment of publicly burning the diamond was successfully made at Florence in 1694, by means of a burning-glass,* when some of the most celebrated scientists attended as witnesses of the important ceremony. The stone, on ex-

* MM. Dumas and Strass, it is said, burned this gem by means of the voltaic battery, an experiment often repeated in modern laboratories.

posure to great heat,* first split, then emitted bright red sparks, and at last was consumed, leaving nothing except carbonic acid gas and a very small quantity of ash. Boyle had previously found that it was combustible, but the correct explanation of the phenomenon belongs to Lavoisier, a French chemist (1743–1794), who not only burned the diamond, but also discovered the true nature of its constituent. Several other experiments were subsequently made establishing the fact of its combustibility, one at Vienna in 1750, and another at Paris in 1771, so that no reasonable doubt in regard to it remains; still, it was thought the diamond might resist the influence of great heat under certain conditions, and an experiment was made in which air was entirely excluded; the stone was neither consumed nor melted, the latter fact proving it to be *infusible.*†

Some experimenters have thought they detected a cellular structure in the residuum, indicating the vegetable origin of this precious stone; but this fact is not, in the opinion of others, well substantiated.

Color.—A large part of the diamonds are white, though a perfectly transparent, colorless gem is more rare than is generally supposed; the remainder present a diversity of hues of different shades, including yellow, red, blue, green, brown, and black. The action of heat in modifying or changing the color has been proved by repeated experiments; in some cases, the original tints are restored after a certain time, but in others the acquired hues are permanent. Though

* It is said that 14° Wedgewood, or five thousand Fahrenheit, is necessary to burn the diamond.

† After the great conflagration at Hamburg in 1842, a large number of diamonds, which had only been defaced by exposure to the heat, were sold for a trifle, under the mistaken idea that they had been permanently injured, but after being repolished they regained their former brilliancy and lustre, with no other loss than a slight reduction in weight.

some of the older mineralogists attempt to account for the different colors of precious stones, their theories are not altogether satisfactory, and until further discoveries are made, we must content ourselves with admiring their beautiful hues without understanding Nature's methods of painting them.

The yellow diamond, perhaps, affords the greatest number of shades, some of them surpassing in beauty every other gem of this color; specimens of a canary tint are quite general. Rose-colored diamonds are not so plentiful as has been supposed; while the red, of rich deep tints surpassing the ruby in beauty, are extremely rare, and constitute one of the most magnificent ornamental stones known to exist. A few of this variety are on record, comprising one weighing ten carats, bought by the Emperor Paul of Russia, for one hundred thousand roubles, and another, referred to by Mr. Streeter, purchased of a London firm by a gentleman in Paris. Several specimens occur of reddish shades, such as garnet, hyacinth, lilac, and peach-blossom, seen in the different collections of Europe. Blue diamonds rank next to the deep red for rarity and beauty, those of a dark blue shade constituting beautiful gems, which differ from the blue sapphire in the quality of the tint, and in the play of colors peculiar to the diamond. The only blue stones known have been found in the old mines of India, none having been discovered, according to this writer (1877), in Brazil or South Africa. He further states that besides the Hope and the Brunswick blue diamonds, there are only three gems of this kind in Europe that can with propriety be called blue, and that all these differ from the Hope and from one another.

Next to the yellow, for colored varieties, the green, including all shades, are most numerous, yet the pure emerald or grass-green diamond is rare, but when it does occur it forms a most beautiful stone, exceeding in brilliancy and fire the finest emeralds. There are several of splendid green tint in the Museum of Natural History, in Paris, but the best known specimen is in Dresden, which Mr. Hamlin considers as "one of the five paragons among all the gems of the world." It has been stated that one of the finest green diamonds yet discovered is owned by an American amateur of precious stones.

Black diamonds are, probably, as rare, or nearly so, as the red shades of this gem. A coal black specimen weighing three hundred and fifty carats, exhibited at the London Exposition of 1851, excited the surprise and admiration of all connoisseurs, on account of its color and size. Those of a brown color, and a variety presenting a cloudy or milky appearance resembling the opal, are sometimes met with.

A fine collection of colored diamonds, gathered by the untiring patience of a Tyrolese, who devoted the greater part of his life in searching for specimens, are seen in the Museum of Vienna. A spray, composed of colored diamonds of all the tints that could be collected in ten years' research, constituted, says Mr. King, the most charming piece of jewelry he ever beheld. The Townshend gems include a great variety of colored diamonds — black, yellow, green, gray, indigo, cinnamon, and others.

Frangibility. — The opinion was formerly entertained that the diamond was infrangible, and could not be broken even by a blow of the hammer, a mistake arising, doubtless, from making hardness synonymous with toughness. The fact is that the hardest of gems is one of the most brittle, and has been broken by simply letting it fall upon the floor. It can be split with a knife in the direction of its cleavage planes, or pulverized in a mortar. Many valuable specimens have been needlessly sacrificed from careless handling, or from not understanding its properties.

Cleavage, an important quality in diamond-cutting, is always parallel to the faces of the octahedron, and, whatever shape it assumes, it can be split into that figure. The magnifying power of this gem is superior to that of glass, which has led to its use, in some instances, for microscopic lenses, though it is said to be difficult to make them perfectly accurate.

A summary of the properties of the diamond is as follows: Hardness, 10; specific gravity, 3.55; cleavage, perfect; refraction, simple; transparent to opaque; combustible; infusible; frangible; phosphorescent; great power of refraction; dispersive power; remarkable lustre and play of colors; positively electric by friction; non-conductor of electricity; does not polarize light.

Classification. — Dieulafait recognizes this precious stone under three different molecular states: First, crystallized — the most usual form, and the one employed in jewelry; second, crystalline, or imperfectly crystallized, as bort, which is excessively hard, far exceeding the ordinary diamond in hardness, and used for powder; third, amorphous or uncrystalline, an opaque, steel-gray mass, called carbonado, used for polishing. It is sometimes of a granular structure, imperfectly crystallized, porous, dense, or massive, with a hardness equal to and even surpassing the crystallized form, and when burned, it leaves a residuum of clay and other substances.

It was formerly believed that the diamond was a kind of rock-crystal, and the latter has sometimes been mistaken for it, but a knowledge of crystallography would have prevented such an error. Quartz crystals are hexagonal, whereas those of the diamond are eight or twelve faced, though the primitive forms are sometimes varied and complex. The normal shape of the Indian specimens is a regular octahedron, the Brazilian are twelve and occasionally six-faced.* Groups of crystals including both forms sometimes occur, as may be seen in a specimen of the Dresden collection. The faces or planes are frequently convex, having been naturally rounded. In the form of its crystals, the diamond may be confounded with the white spinel.

Microscopic cavities or fissures exist in many specimens, which give them a dark color, while others present a stellated appearance.

Sometimes this gem will burst or split from natural causes, a singular phenomenon happening with certain glassy stones of a faint brown tinge. Specimens perfect when taken from the mines have been known to be lying in fragments the next morning. It has been suggested that this catastrophe is occasioned by the vaporization of the water between the laminæ, induced by atmospheric heat, but the real cause is a question for speculation.

Supernatural powers have been ascribed to the diamond, as if its natural properties were not sufficient to constitute it one of the most remarkable substances in nature. There has been a difference of opinion about its medicinal qualities — some believing it was a deadly poison if taken into the stomach, while others have regarded it an antidote to poison. It is said that the life of Benvenuto Cellini was attempted by one of his rivals, by administering to him a draught containing, as was supposed, pulverized diamond, believed to be a virulent poison, but the danger was averted by the cupidity of the apothecary, who prepared the deadly beverage by substituting the beryl, a cheaper gem. The mysterious death of Sir

* The cube form of the diamond and the hard round bort, the latter being really a twinning of the cube, are peculiar to Brazil. — *G. F. Kunz.*

Thomas Overbury in the Tower of London, at the beginning of the seventeenth century, was ascribed to a potion given to him by an enemy, containing a preparation of the diamond.

Uses. — A difference of opinion has prevailed among nations and individuals as to the rank this gem is entitled to among precious stones. The ancient Romans and the people of India assigned it the highest place for beauty and value, while the Persians esteemed it less, giving it only a fifth rank, the pearl, ruby, emerald, and chrysolite taking precedence; other nations have considered it inferior only to the ruby and the emerald, though the majority of mankind have regarded the diamond as the queen of gems and the nonpareil of all material things, whose possession was once claimed as a regal privilege and none except those of distinguished rank presumed to appropriate it for personal ornament; but now the imperial gem has become the legitimate property of any one who can purchase it.

The Syrians and Phœnicians are supposed to have been the first nations who employed it for jewelry, and their example was soon imitated by others. At a later period it was introduced into Europe and became conspicuous at all the courts and in all the circles of rank and fashion, while at the present time it maintains a pre-eminent distinction as an ornamental stone in both hemispheres.

The French collection of diamonds was large at the time of the great robbery, and since that affair, accessions were made from time to time until the list reached, in 1838, the enormous sum of nearly sixty-five thousand specimens, including many of distinguished size and beauty. Additions continued to be made to the collection until the number was almost without a parallel.

In 1872 the Buonaparte family alone, within one year,

threw upon the market diamonds amounting to one million two hundred and ten thousand dollars.

A magnificent bouquet of French diamonds, so ingeniously constructed that it could be taken to pieces even to the petals of the flowers and converted into seven different brooches, was exhibited at the London Exposition in 1851.

The late Duke of Brunswick, who had a passion for the acquisition of precious stones, left at his death a collection of diamonds alone, estimated at two million five hundred thousand dollars, but his possessions gave him little happiness, since he lived in continual fear of being robbed and murdered, and felt compelled to take annoying precautions against such an event. Though these treasures may have been the source of danger in numerous instances, and the incentive to terrible crimes, yet on one occasion, at least, they afforded a protection to life. It is said that when Isabella II., Queen of Spain, was attacked in public by a bold assassin, the dagger of her assailant was intercepted by a diamond girdle worn by this princess, and before another blow could be given, she was rescued from her perilous condition by her attendants.

As an example of the abundant use of the diamond in earlier times, it has been estimated that Sultan Mahmoud left at his demise, at the beginning of the eleventh century, more than four hundred pounds, avoirdupois, in these gems.

The ancient Romans and early mediæval races valued this precious stone more on account of its credited supernatural powers, than for its intrinsic beauty, which had not been developed by the later process of faceting. It was early selected for the marriage ring, on account of its power to promote harmony in the conjugal relations. Those earliest known to the Romans were obtained from Ethiopia, but only a few of their diamonds have any historical interest like those of mediæval and modern times. One of these gems figures in the chronicles of the empire as having been owned by the Emperor Nero and transmitted to his successors, Trajan and Hadrian.

The story of the "Diamond Necklace" has often been told, but it will bear repeating, like many another tale fraught with romantic interest. This jewel is said to have played a prominent part in bringing about the French Revolution, in which Marie Antoinette, Cardinal Rohan, Madame De La Motte and her husband, are the principal actors. Briefly stated, the facts are these: Louis XV., in 1774, commissioned two court jewellers to make a necklace of the most beautiful diamonds to be had, for Madame du Barri; but before it was completed, the king died and was succeeded by Louis XVI. The necklace was, however, finished, with the hope that it might be purchased for Marie Antoinette, the new queen; but the price, between one and two million livres, was beyond the capacity of the "royal exchequer." Subsequently, Madame De La Motte, one of the queen's attendants, represented to Cardinal Rohan that her majesty had reconsidered the question and would enter upon negotiations for the purchase of the necklace. Duped by this woman, the cardinal bought the jewel, at the request of his sovereign, as he supposed, and consigned it to the attendant, whose aim was to get possession of it. De La Motte, the husband, escaped with it to England, where it was broken up, with the view of disposing of the diamonds. In the meantime, the jeweller, who believed the queen had been the purchaser, brought his claims for indemnity; consequently, the plot was discovered. The unfortunate cardinal, whose only fault in the matter was over-credulity, was sent to prison, and Madame De La Motte was scourged and condemned to perpetual imprisonment, from which, however, she managed to escape. This fraud, one of the most daring recorded in history, excited public interest throughout Europe, and caused a decline in the use of the diamond in France; and during the Revolution, which soon followed, it was entirely ignored.

Artificial Diamonds.— Attempts have been made to produce these gems by an artificial process, but thus far the efforts have not been attended with very gratifying results. M. Despretz, after repeated experiments made in 1828, succeeded in obtaining some minute crystals resembling diamonds, which, however, were regarded as failures.* These abortive results did not deter sanguine experimenters from making further attempts to discover a method of manufacturing diamonds, or a production so closely resembling them as to replace the genuine article, from boron, one of the elementary substances, which closely resembles carbon, the constituent of the diamond, in several of its properties, while its transparency, power of refraction, hardness, play of colors, and resistance to the action of nearly all chemicals, render it a desirable substance from which a gem of such excellence as to compete with the diamond might be artificially formed.

On the other hand, it must be stated that boron crystals are very difficult to be obtained, and when secured, they have been thus far only of very small size. Glass imitations are produced with much greater facility, and they closely resemble the genuine diamond in brilliancy and prismatic effect, though they

* At a meeting of the Royal Society in 1879, a paper was read by Mr. Hanny, entitled "On the Solubility of Solids in Gases," which stated some interesting experiments relating to this subject. It was shown that a crystal of potassic iodide could be dissolved in alcohol gas, and become again crystallized. The experiment was successfully tried with various solids. This suggested the idea that if some gaseous solvent could be found to dissolve carbon, artificial diamonds could be produced. Mr. Hanny succeeded in finding such a solvent, and obtained some minute crystals containing 97 per cent of carbon, and with all the attributes of natural diamonds.

are inferior in hardness, and wanting in the adamantine lustre, as well as some other characteristics of this precious stone. An attempt has been made to give yellow diamonds a blue tint by means of aniline dye or a certain kind of blue pencil, with a view of enhancing their commercial value, but the fraud is too palpable easily to escape detection. The latter process consists in wrapping the stone in a damp sheet of tissue paper which has been rubbed with powder abraded from a blue pencil, and to prevent detection, the gem is set before it is offered for sale.

Counterfeit Diamonds. — There are so many colorless gems which resemble the diamond that it is quite difficult, if not impossible, for one, not an expert, to detect the spurious from the genuine article. Rock-crystal, colorless varieties of spinel, topaz, emerald, sapphire, beryl, and zircon, have all passed for diamonds. White zircon and phenakite, of all natural counterfeits, come nearest to them in play of colors; rock-crystal is inferior in lustre. The Novas Minas or white topaz of Brazil, sometimes called the "Slave's diamond," is very hard and brilliant, but lacks the adamantine lustre and iridescence of the true diamond.

Various methods are employed to detect counterfeits of this gem, depending upon its peculiar properties, such as its electrical powers, its single refraction by which it is distinguished from most other precious stones; but the most decided tests are, probably, those of hardness and specific gravity. The most convenient method of testing the diamond, it has been said, is to submit it to a white heat and then apply the point of a sapphire. If it is genuine, it will undergo the former ordeal without melting, and the latter without being scratched.

Mines and Mining. — The uncertainty and hazards attending diamond-mining render it to a great extent a gambling operation. The business is attended with immense labor and expense, and sometimes with great risks, making the profits, at best, very precarious. The methods of working the mines, called the "wet" and the "dry" processes, are similar in different countries; therefore a description of the Brazilian operations will give a pretty correct idea of those in other regions. At the time of Mawe's visit to these mines, during the early part of the present century, the diamonds were usually obtained from the beds of rivers and deep ravines, though they were occasionally found in cavities and water-courses on the summits of the loftiest mountains.

By turning the streams from their natural courses, the diamonds deposited in the gravel (called *cascalho*) constituting the beds were obtained by a process of washing the sand for the large specimens, after which the pulverized earth was placed on the "depositing ground," and worked over for the smaller stones. One river alone, according to this traveller, had been a fruitful source of supply for half a century before his visit.

"The dry method of working the mines," writes Burton, whose travels were more recent than those of Mawe, "long known in India, is not practised in Brazil," but the manner of "washing" for the gems as described by him was that similar to the oriental, and pretty much the same as Mawe relates it. Numerous ablutions were necessary before all the diamonds were assorted, and it required from half an hour to one hour for a good washer to finish one panful of cascalho. The work is said to be very severe for the eyesight, which, in a few years, is greatly impaired; therefore, children, whose vision is more keen than that of adults, are considered the best washers.

The Brazilian and Indian process of washing for the diamonds has been introduced into the South African mines, with very satisfactory results in saving numerous specimens of small size, which constitute a large per cent of all the diamonds found in this region.

As an illustration of the labor required in this business, it is stated that four hundred slaves were employed three months to remove a heap of cascalho estimated at fifty thousand dollars.

As an inducement to diligence and honesty, freedom was offered to every slave working in the mines who found a diamond weighing seventeen and one-half ounces. This event was attended with considerable ceremony: the fortunate discoverer was crowned with a wreath of flowers, and carried in procession to the superintendent, who gave him a new suit of clothes and his liberty, together with permission to work in the mines on his own account. A touching incident is mentioned of a slave who just missed the boon of freedom by the lack of only one carat.

Notwithstanding premiums were offered for large stones, and penalties in the form of chastisements and imprisonments were enforced for purloining the diamonds, yet a great many were secreted by diggers even under the strict watchfulness of the directors.

Some of the South American mines were remarkably productive, yielding in a single locality from twenty thousand to twenty-five thousand carats annually. The Rio Pardo, an insignificant stream, afforded large quantities of bluish-green stones, and the Valho those of large size and great brilliancy. The diamond region of Cerro do Frio is said to cover about fifty miles in length and twenty in width. When first discovered, the diamonds of this region were regarded as worthless pebbles, until some of them found their way to Lisbon, thence to Amsterdam, where their true nature was recognized.

The Brazilian government immediately took possession of the territory, and assumed control of the mines. The amount of this precious commodity exceeded the expectations of the most sanguine officers of the crown, and was sufficient to meet the demands of the Indian and European markets for a long time. It has been estimated that during the first half century after their discovery the value of the diamonds exported from the mines of Brazil reached the sum of sixty million dollars. The greater part, however, were small or of moderate size, not more than one specimen in one hundred thousand weighing thirty-six or more carats. The exceeding richness of the mines stimulated private parties to ask for the right to work them on their own account. This privilege, granted by the government, soon led to all kinds of frauds, so that it was compelled to resume control of the business in 1772; consequently, all the diamonds after that date have belonged to the crown. The most valuable gems exceeding seventeen carats have been appropriated by the royal family, and, for this reason, the imperial treasury, since the beginning of the present century, has nearly if not quite equalled that of any other royal collection in the world, for the number, size, and quality of its diamonds. Their aggregate value has been estimated at several million dollars.

The diamond mines of Borneo are among the oldest and most productive in the world; as early as 1738, says Mr. Streeter, the Dutch were extensively engaged in the business, and annually exported from this island between two hundred and three hundred thousand dollars worth of diamonds. There were few European courts of that period which could vie with the Batavian in the rich and brilliant display of these gems.

The diamonds of the Borneo mines occur in beds from ten to thirty feet in depth, at the foot of mountains, the largest and best specimens being found in the lowest strata. The Chinese worked many of these mines until the middle of the present century, when they were driven away by the natives, a measure which was immediately followed by a decline in the supply.

Diamond-cutting. — Having secured these valuable treasures at great expense and trouble, the question is pertinent, what shall be done with them? If the fortunate possessor is only a connoisseur, he will arrange them in a cabinet for the admiration of himself and friends, but if he is fond of personal ornaments, he will want them cut and polished, then mounted in a convenient form for use, since in their native state they are not adapted for jewelry, and need the skill of the artist to develop their inherent beauties.

The practice of cutting and polishing gems is not as modern as some writers are inclined to believe, but was known very early in the history of art, if Hill and others are correct in their statements. It is thought the Phœnicians learned the process of cutting precious stones from the Assyrians, and soon diffused it through all their colonies, and that cutting and mounting them were understood in Great Britain during the Roman period. It is not positively known that the diamond was included in their list of gems; neither is it certain it was not.

We meet with conflicting statements in regard to the origin of the modern style of cutting and polishing diamonds. It has frequently been ascribed to Louis de Berghem or Berquem, of Bruges, of the fifteenth century, and that the first diamond he cut was for Charles the Bold, Duke of Burgundy. On the other hand, it is maintained that the art of cutting this gem was known long before his day, as is proved by the existence of very ancient church ornaments cut as four-sided pyramids, and by the cut diamonds mentioned in the inventory of the jewels of the Duke of Anjou, in the fourteenth century. Probably the truth lies between these opposite statements; that though the lapidary of Bruges did not invent the art of cutting, he introduced many improvements by the use of the polishing wheel and by facets cut according to mathematical principles, which placed the business among the sciences as well as the arts.

Amsterdam has held the foremost position for diamond-cutting, though it has had its rivals in other places. A society of lapidaries was formed in Paris the last of the thirteenth century, and during the supremacy of Cardinal Mazarin, it held an important place in the circle of the arts; but the business finally passed to Amsterdam, where it maintained its supremacy for a long period. Mr. Costar's establishment in that city is claimed to be the largest of the kind in the world. It has been estimated that more than fifteen-sixteenths of all the diamonds cut and polished, at least to a very recent date, have passed through the hands of the Amsterdam lapidaries, of whom there are ten thousand constantly engaged in this occupation. During the last century, the chief seat of the business was London, and those cut there commanded a higher price than those of the modern Dutch finish; but this city could not compete with her rival, and the industry declined until the discoveries in South Africa, which gave it a fresh impetus. The opinion has been expressed that the best cutting and polishing in Europe, at the present day, are done in Paris; while London is the greatest market for rough diamonds from all the different mines.

Diamond-cutting is performed very successfully in the United States by Messrs. Tiffany and Company, New York, whose establishment has been described in another chapter, and by Mr. H. D. Morse of Boston, whose genius, says Mr. Hamlin, led him to invent a machine for cutting and polishing gems so that American jewellers could have their work done at home. He also succeeded in educating a corps of native workmen, thus avoiding the necessity of importing cutters from Amsterdam. Mr. Morse has cut and polished many large gems, including one from the South African mines weighing one hundred and twenty carats, reduced by the operation, which required between three and four months' labor, to a beautiful gem of seventy-seven carats.

Some interesting facts about diamond-cutting were given at a meeting of the New York Academy of Sciences, April, 1885, by Mr. G. F. Kunz, when he stated that an experiment was made by Messrs. Tiffany and Company upon a diamond from Brazil, a variety composed of numerous twinnings, showing its extreme hardness. The stone was placed on a polishing-wheel, with a circumference of two and one-half feet, the wheel making 2800 revolutions per minute. Besides the weight of the holder, usually less than three pounds, additional weights were added from time to time, varying from four to forty pounds, causing scintillations to be thrown off, and ploughing the wheel, rendering it unfit for use. This process was repeated for one hundred days, yet such was the intense hardness of the diamond that it had received scarcely any perceptible polish.

It may be of some interest to the amateur or the owner of diamonds to have a general idea of the method of cutting and mounting them, as their beauty and commercial value are more or less affected by the manner in which they are dressed. The skilful lapidary will observe certain proportions between the several parts of the stone; otherwise there will be a sacrifice of brilliancy. This fact is illustrated by the Pitt or Regent diamond, considered the most faultless brilliant known, and the Koh-i-noor, in which proportion is sacrificed to save loss of weight,* the breadth being too great for the depth, consequently it is deficient in brilliancy.

The forms of cutting precious stones vary; they are generally classed as table, rose, brilliant, brilliolette, step, and cabochon. What is called step-cut is adapted for many of the transparent, colored stones, while the translucent and the opaque varieties are usually cut *en cabochon* — that is, without facets, or as convex, concave, double convex or having one flattened and one convex surface. The garnet, it is claimed, is the only transparent gem cut to advantage *en cabochon;* others, like the ruby, sapphire, and zircon, lose in brilliancy by this method.

All diamonds, several centuries ago, were cut with a square or oblong plane on both sides, one being much smaller than the other; they were designated *table* or Indian cut. At a later period a form called the *rose* came into fashion, which consisted of a flat base and a dome above, usually with a double row of facets presenting a figure like a half polyhedron. The rose is much less expensive than the brilliant cut, and can be fashioned out of very flat or cleavage stones. It has been stated that rose-cut diamonds are sometimes of a size so small as to require fifteen hundred to weigh one carat, which seems incredible when the extreme delicacy and skill required for such work are considered. This style of cutting receives different names, according to the number of facets the stones display.

The *brilliant* is a later invention, and the one most in use; its origin has been referred to Peruzzi, of Venice, of the seven-

* As a rule, the diamond loses from one-third to two-thirds of its weight by cutting and polishing.

teenth century. It has been described as presenting the appearance of two cones united at their base, the upper being truncated. The technical terms for the different parts are the table or flat upper surface, called the crown, the pavilion or base, comprising the lower part of the stone, the culet or collet, the under plane opposite the table, the girdle or the junction of the pyramids, and the beasil, or slanting edge.

There are certain proportions to be observed in this mode of cutting, as the table must be four-ninths the size of the stone, the collet one-sixth of the table, from the table to the girdle must be one-third of the whole thickness, and from the girdle to the collet two-thirds. The perfect brilliant requires at least fifty-six facets, thirty-two above and twenty-four below the girdle; sometimes they have more. These facets are of various forms and sizes, designated by different names, as star facets, skill facets, and others. The above rules are not always strictly observed in cutting.

The *brilliolette* has been described as two rose-diamonds entirely covered with small facets, and joined at the base: several notable diamonds are cut in this style. The point, a name still in use, consists of a four-sided pyramid. A diamond should be cut with a thin edge at the girdle in order to display its prismatic play of colors to the best advantage. Experienced lapidaries regard Indian-cut stones and many of recent workmanship defective in their style of cutting: those from the East are frequently produced from flat, veiny stones, called lasques, in the form of single brilliants, which are greatly inferior to the double brilliant.

It has been thought the art of cutting the diamond originated in India, though the natives prefer the gem in its rough state or as polished by a natural process, when they are called "naifs." Oriental lapidaries resort to the skilful artifice of covering a defective specimen with facets, in order to conceal its imperfections.

The ancients may have used the diamond-point for cutting and filling in the details. It is certain they used the corundum for the softer gems, a practice introduced into Greece from the East. In modern finish, the work is accomplished by a revolving disc and diamond powder, or carbonado; whence the metaphorical expression, "diamond cut diamond." The earlier Gnostic gems were worked by the diamond-point, while the later specimens were submitted to the wheel, so that an expert can easily distinguish an ancient intaglio from a modern imitation by the method used for cutting it. The stone to be operated upon is first split, and, as it consists of a series of layers, this can be accomplished in the direction of these laminæ; the next step is to cut them, which is accomplished by rubbing together two diamonds of nearly equal size, cemented into a handle; the powder made by erosion is preserved for polishing. This work is performed upon a disc of soft iron, made to revolve very rapidly upon a horizontal plane covered with powder mixed with the purest olive oil. When one facet is completed, the stone is changed in its position, so as to present another side until it is completed; more than one stone can be worked at the same time. The process of cutting requires the greatest care and nicety, since the least carelessness may spoil the gem. According to Emanuel, the finest specimens are sent to the London market, those of a second rank to Paris, and the inferior stones are exported to the United States, Turkey, and other countries. If this is correct, London is the best place for the purchase of these gems, where quality is the first consideration; but it has been stated, on the other hand, that the most costly finished diamonds find sale not only in England, but also in France and the United States, while those of inferior grade are purchased in eastern Europe.

Gold is most frequently used for mounting gems, though silver is considered by some lapidaries the most appropriate for colorless varieties, since it preserves their transparency more effectually, and enhances their brilliancy. One method of mounting is by the close setting, with only the upper part visible, and another by the open setting, leaving the edge of the stone clear: the latter is preferred for the diamond.

Engraved Diamonds.—It seems strange there should be any difference of opinion on this subject among writers on precious stones, but such is the fact; some who ought to know maintain that it neither has been nor can be engraved, while others support the opposite view. The historical evidence that engraved diamonds have been known, seems to be conclusive. Corsi, an Italian antiquary, says Ambrose Caradossa, in 1502, was the first to engrave the diamond, and accomplished a work of the kind for Pope Julius II. The discovery of the method has been ascribed by others to Trezzo, a celebrated Milanese artist, who executed several engravings upon this gem. His first attempt represented the coat-of-arms of the Emperor Charles V.; then followed the portraits of Don Carlos, Mary Queen of England, Mary Queen of Scots, and the arms of Phillip II. of Spain, all the works of this engraver or some of his pupils. A number of intagli were cut on diamond by the Cinque-cento artists. According to Stosch, both Giovanni and Carlo Costanza, modern Italian artists, executed some fine portraits on the diamond, mentioned in the chapter on "Engraving on Precious Stones." Other instances could be cited, but they are not needed to prove what is so well known. Several engraved diamonds, the productions of the Costanzi, were stolen at the robbery of the Galleria della Gemma, in Florence, in 1860.

The royal collection of England is said to contain the diamond signet-ring of Charles II. when Prince of Wales, bearing the ostrich plumes.

Trade and price. — Diamonds are subject to the same commercial laws which govern the value of every other commodity — those of demand and supply. There has been a gradual advance in the price of these gems for the last thirty years, partly on account of the exhaustion of the Indian mines, partly the falling off in the supply from Brazil, and partly on account of the constantly increasing demand for them. The proportionate increase in the price for small stones has been greater than for large ones. Colored diamonds, when of a decided, beautiful hue, as red, green, blue, and some other tints, frequently bring a very great price. It is said that a fine green specimen of only eight grains, belonging to the collection of the Marquis of Drèe, was sold for a sum equal to a large fortune. It has been estimated that diamonds represent ninety per cent of all the large amount of capital invested in precious stones, which proves their use as an ornament vastly exceeds that of any other gem. They are sold by a weight called a carat, a word of Indian origin, supposed to be derived from the seed of a plant, and varies in different countries; formerly, it was reckoned at four grains troy, "even beam," but now it falls below that weight.*

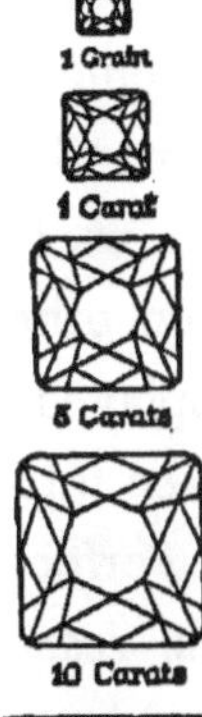

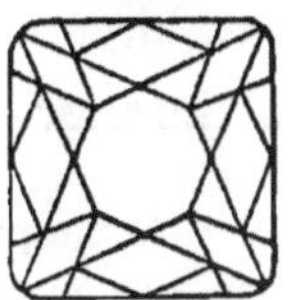

Different sizes of brilliants.

In its early history, the diamond had no fixed standard of prices, but later a rule was adopted, by which the value was

* It is reported that a standard of the diamond carat at .205 grains was agreed upon by a syndicate of Parisian jewellers, goldsmiths, and gem dealers, in 1871, and was subsequently confirmed by an arrangement between the diamond merchants of London, Paris, and Amsterdam.

reckoned in proportion to the square of its weight; or the value was the square of the weight in carats multiplied by eight;* as, for example, a diamond of one carat being forty dollars, according to this computation, one of two carats would be one hundred and sixty dollars, and one of ten carats four thousand dollars. This method of valuation, which is not now used, could be applied only to stones of moderate size, since those of great weight, sold in this way, would cost a fabulous price. For more than a century, the value of the diamond has been based upon the form of the brilliant, while that of the table, rose, and other styles of cutting depends upon other circumstances. The tint, the water, and the skill displayed in cutting and polishing are considerations to be taken into account, when purchasing this gem.

Perfectly colorless diamonds, entirely free from all impurities, resembling a drop of the clearest water, and exhibiting the highest lustre, are said to be of the "first water," those of inferior grade of the second, etc.

* At the present date, a diamond of one carat, and of perfect form, is worth from one hundred to two hundred dollars.

CHAPTER XI.

HOME OF THE DIAMOND.

India.—This country, the source of a large part of the luxuries of the East, and the home, par excellence, of the most valuable of the precious stones, has always supplied the rest of the world, to a great extent, with its ornaments; consequently, it became the great mart for the traffic in gems. The diamond employed for personal decoration by the ancient people of this venerable empire, as we learn from their literature, must have been an article of commerce among the Asiatics at a very early period in the world's history, though the first authentic account of the diamond mines of India, by a European, is comparatively modern. Garcias, a Portuguese traveller, first made known their existence, in 1565, but the first published record was made by Jean Baptiste Tavernier, a French jeweller and famous traveller of the seventeenth century, who visited four different mines, which he describes at some length. The Raolconda mine,—not Golconda, as is sometimes said; there are no diamond mines there,—was discovered more than four hundred years ago. The soil was found by this explorer to be full of rocks crossed by fissures enclosing sand with diamonds, but they were often flawed, a defect which the native lapidaries ingeniously concealed by facets, and, remarks this shrewd observer, a stone dressed in this manner was pretty sure to be imperfect. The cutting was accomplished with great facility, but the polishing was inferior to European finish.

Very young persons were often employed in some departments of the business, and soon acquired the habit of judging the value of a stone with remarkable discernment. Rewards were offered for all specimens exceeding fourteen carats weight, but this inducement was powerless in preventing frequent thefts; consequently, some of the best specimens were appropriated by the miners. All diamonds above ten carats in weight were reserved for the royal treasury. The Raolconda mines were at that time considered the richest in India, if not in the world, and employed thirty thousand laborers at once to work them.

The Coulour, Colore, or Gani mine, the former name received from the Persians, the latter from the natives, was discovered about a century after the Raolconda; both are in the south central part of India, and several days journey from Golconda. The Gani diamonds were accidentally discovered while digging a piece of ground for agricultural purposes, by a native laborer, whose first prize was a stone of twenty-five carats, soon followed by a plentiful harvest, which yielded some enormous gems, including the Great Mogul, and another, weighing nine hundred carats, presented to the Emperor Aurungzeeb. Though the Coulour or Gani mines were remarkable for the number and size of the stones, yet these were not generally of the purest water, many of them being tinged with green or yellow.

There were about sixty thousand persons, including all ages and both sexes, engaged in these mines at the time of Tavernier's first visit.

The diamond localities in the region of Raolconda were once numerous; this traveller mentions as many as twenty mines, but since his day they have all been abandoned except two or three. The Punnah beds, in north central India, situated on a table-land elevated from twelve hundred to thirteen hundred feet above sea-level, have been famous since the reign of the Ptolemies. This region, visited at the beginning of the present century, was found to contain diamonds in gravel beds at a depth of from six to twenty-four feet.

The Sumbulpoor mines, in the region of the Mahanuddy river, are of great antiquity, and supposed to be the oldest in India. At the time of Tavernier's visit, some of the mines in the Carnatic province had been closed by order of the emperor, on account of the imperfection of the gems. Indian diamonds were obtained from river-beds by a process similar to that practised in Brazil.

Diamonds have been found in numerous localities in India, besides those just designated. Mines were formerly worked along the Coromandel coast, and among the hills of Bengal, but they have also been abandoned, from one cause or another — it could not have been from exhaustion; and this country, at the present time, not only furnishes none for the market, but depends upon exportations for her own supply; still, it is thought there is an abundant store of these precious stones in this wonderful land, which may yet be developed.

Borneo. — This extensive island is supposed to be very productive in diamonds, but authentic accounts of the mines are very meagre, owing to the obstacles in the way of exploring, arising from the opposition of the natives, and the difficulty of penetrating the interior, where they occur. The character of the mines, as far as is known, is similar to those in India, the gems being found at different depths in gravel. The diamonds of Borneo, said to be the best in the world, are celebrated for their remarkable adamantine lustre.

Brazil. — It was not until 1727 that the real character of the Brazilian diamond was known; this was obtained not from the natives, — they were ignorant of the gem in its original state, — but through a Portuguese, who, suspecting their true nature, sent some diamond pebbles to Lisbon for examination. The tests applied proved his suspicions well founded, and the discovery created a great sensation among the dealers in Europe.

The Dutch, who had a monopoly of the India trade, made an effort to depreciate the Brazilian diamonds by pronouncing them spurious, but they were countervailed in their designs by a skilful manœuvre of the dealers, who first sent them to India, then reshipped them to Europe as Indian stones, and had them cut after the Indian fashion.

Stories about the productiveness of the South American mines seem fabulous. The diamonds were found scattered about in the most lavish manner, and were picked up by children and slaves, and even seen adhering to the roots of vegetables. The vast extent of these diamond-fields, and their exceeding richness in this precious stone, at first caused a great panic in commercial circles, from fear of overproduction, but that result has been greatly neutralized by the difficulty and danger of working the mines, and the constantly increasing demand for this gem. As an instance of the immense yield in Brazil, the Bank of Lisbon is said to have sold, in 1863, a collection of these diamonds made by John VI., in 1821, valued at one million eight hundred thousand francs, and still there remains to the Portuguese crown an overplus, in these gems, valued at thirty-five million francs. Europe received from Brazil, during the first twenty years after the discovery of the mines, more than three million carats of diamonds. The total production from 1861 to 1867 was nearly one and one-half million dollars.

The various estimates in regard to the yield of the Brazilian mines are, no doubt, to a greater or less extent, conjectural, but there remains no uncertainty about their unparalleled wealth when first opened. Though the production has been decreasing for some time, yet it is believed that a vast extent of diamond-bearing territory remains to be explored.

These mines afford, as a rule, stones of much smaller size than those of India, not more than one in ten thousand exceeding twenty carats; yet there have been some specimens of great size, including the Star of the South, the Braganza (if it be a diamond), and several others of less weight. The Portuguese government declared, in 1730, that the diamonds found in the regions lying between twenty and thirteen degrees south latitude, belonged to the crown; these boundary lines were called the "Diamantina demarcation."

Mawe visited the mines of Brazil in 1812, a favor never before granted to a foreigner; no Portuguese, even, had ever been allowed to approach them except on business. This traveller relates an amusing story about the discovery of a diamond of immense size, in this wise: A free negro of Villa do Principe informed the Prince Regent that he was in possession of a very large specimen, and begged the honor of presenting it in person to His Royal Highness. Accordingly, an escort was sent to bring the diamond and its owner to the capital, a journey which occupied twenty-eight days, and was the occasion of great public interest. Much surprise was awakened on its arrival, that so large a diamond could possibly exist; it weighed nearly one pound. However, it was sent to the treasury under a strong guard, and the English traveller was requested to give his opinion of its merits. The test of scratching with a diamond was applied, when, greatly to the disappointment and chagrin of the parties concerned, it yielded to the friction, — it proved to be rock-crystal. Instead of a magnificent fortune and distinguished honors, the luckless discoverer received no compensation for his trouble, and was left to find his way back as best he could.

The richest mines, in the opinion of King, are those of the Sierra da Frio, which, since their opening in 1727, have yielded more than two tons of diamonds. Burton, who visited the mines of Minas Geraes, says, since their opening in the seventeenth century, to 1850, they have yielded nearly six million carats, valued at more than fifty million dollars, besides those surreptitiously secured by the miners. In a single year the Portuguese imported from these mines between eleven hundred and twelve hundred ounces of diamonds. The government, to protect its assumed right to all the gems found in the territory, had recourse to the unjust act of driving away the inhabitants living on the banks of the rivers where they were found, and many of the poor fugitives perished from want, before the edict for their restoration to their possessions was promulgated, in 1805.

The Bahia mines, embracing a territory eighty miles long and forty wide, which were discovered by a slave, were opened about forty years ago, and to the year 1880 had yielded nearly ten million carats of diamonds. The production was so great that their value was reduced at least one half, but at present the yield is considerably less. The stones are found in the cascalho taken from the beds of streams, and sent to Rio, at great trouble and expense, for exportation to foreign markets. The South American diamonds are known in commerce as the Diamantina and the Cincora ; the former are considered the best.

The Brazilian diamonds were formerly thought to be different in some of their essential qualities from the Indian, but it is now conceded that they are alike with the exception of a slight difference in specific gravity, the oriental being a little heavier.

The largest part of the South American diamonds are colorless, while the remainder display a great variety of hues,—blue, green, yellow, brown, pink, milky white, and, rarely, black. The collection of the Prince Regent, at the time of Mawe's visit, comprised a specimen exhibiting several colors, a rare occurrence with this gem.

South Africa. — The discovery of the diamond-fields of this region, in 1867, created a great sensation in both hemispheres, and awakened a spirit of enterprise among those who were eager to amass a sudden fortune. They occupy a small portion of the western part of the Orange River Free States, established by the Dutch about the middle of the present century, covering the territory between the Orange and the Vaal rivers, extending, however, into some of the neighboring states. Although a surprise to foreigners, the diamonds had been known to the natives long before, and, as early as 1750, were employed for drilling rocks.

The recent discovery was accidentally made by a trader, on one of his journeys into the interior, who, while passing the night with a Dutch farmer, had his attention attracted to a pebble with which the children were playing, on account of its resemblance to a diamond. He expressed his opinion to his host, and upon examination by a competent judge, it proved to be correct; the pebble was indeed a genuine diamond, weighing twenty-two and one-half carats, and was subsequently sold for several thousand dollars. Not long after this discovery the "Star of South Africa" was found, which brought fifty-six thousand dollars.

New and rich fields were frequently developed during the years 1870 and 1871, which stimulated the government of Great Britain to take possession of the whole territory by formal proclamation, an act which set aside the treaty making it a free state.

These diamond-mines are of two kinds: they are known by the name of "river diggings," existing in deposits of gravel containing agate and jasper pebbles; and "dry diggings," in rocks not disintegrated. The bed of the Vaal, as well as the drift on its banks, says Morton, is made up of agate, jasper, chalcedony pebbles, and quartz crystals, with diamonds interspersed. Judging from the appearance of these gems found in the gravel, it is thought they must have travelled far and perhaps been subject to glacial action.

The mines called the "dry diggings" are found in what is called the Karoo formation, a conglomerate occupying an extensive plateau, elevated some five thousand feet and spread over a territory covering two hundred thousand square miles. The diamond was not, probably, formed in the place where it was discovered, but is an accidental constituent of the pudding-stone. Fragments of the broken crystals, as well as entire pebbles, are scattered with remarkable regularity throughout the mass, but are never found near one another. It is mentioned as a rare occurrence that a diamond geode was brought to light during the mining at the Cape.

The soil directly below the surface is described as chalky, and interspersed with nodules enclosing diamonds. Under the chalk rests a yellowish mass, containing more of the precious substance, but the richest harvest is gathered from the conglomerate called by the miners "blue stuff," found at a depth of fifty or sixty feet, and often inclosing topaz, zircon, jasper, agate, opal, and other precious stones. A claim comprising three hundred square feet of this conglomerate has been estimated worth from five thousand to forty thousand dollars, — the difference of price varying according to circumstances. At a place called Du Toit's Pan, twenty-five miles from the "river diggings," the mines, covering many acres, have yielded several large diamonds, the largest reaching nearly three hundred carats. The discovery of the diamonds of this region was purely accidental and due to a Dutch farmer who first noticed them imbedded in the mud walls of his cottage, which led to an examination of the surrounding soil and a knowledge of the wonderful treasures it inclosed.

The New Rush, or Kimberly mine (referring to Morton), is the site of more natural wealth than any other known spot on the globe; it has yielded an immense quantity of diamonds, including some of remarkable size, as the Stewart, and three others, named by Dieulafait the Koh-i-noor of South Africa, the Star of Beaufort, and the Star of Diamonds. Information has been received that the Kimberly mine has suffered a serious, if not a permanent injury, by a caving-in of its walls.

The Cape diamonds, as the South African specimens are called, are unique in some of their characteristics: they have no skin or envelope, like the Brazilian. The mines do not yield carbonado, but sometimes the crystals group themselves about the kind called bort, and form a mass of natural brilliants. Those obtained from the beds of streams are usually white, and of a quality superior to the tinted.

The colors of the African diamonds range through all shades of orange and yellow to faint straw, and occasionally they are blue, brown, pink, green, and black; the white or colorless specimens are said to be as fine as any from India or Brazil. It has been estimated that ten per cent of the stones are of the first quality; Mr. Streeter reckons twenty per cent, and fifty-five per cent fit for diamond-cutting.

The total yield from their discovery to 1876 has been rated at eighty-five million dollars. Many fortunes were made in these mines, though none enormously large, rarely above fifty thousand dollars, a significant fact indicating the great expense and uncertainty of the business. The extent and resources of these diamond-fields are still problems, but as far as they have been explored, there is little to fear from exhaustion for many generations to come; while, on the other hand, there is small probability that the yield will ever be so enormous as to degrade this peerless gem to a common rank among ornamental stones. Nature, it has been said, has guarded the diamond with special care, by placing it in regions difficult of access, or in situations taxing the ingenuity and endurance of man to get possession of it.

Australia. — The discovery of Australian diamonds was made as early as 1860, but operations for mining did not begin until nine years later, under the management of the Australian Diamond Miners' Company; the principal mines are the Bingera and the Mudgee. The mines of the Bingera district, four hundred miles north of Sidney, on the Big River, were discovered in 1867, by gold diggers, as it generally happens the diamond is found with or near this precious metal; no stones were found in the rivers except where the soil from the gold washings had been discharged. The Bingera diamonds occur either in the Devonian or the Carboniferous strata, scattered from a few feet below the surface to nearly seventy feet in depth.

The Australian diamonds are sparsely scattered, and are generally of a small size, the largest not often surpassing eight carats.

The first diamond found in Australia, a stone weighing three-fourths of a carat, is exhibited in the Museum of Practical Geology, London.

Russia. — In the early part of the present century, the attention of geologists was directed to eastern Russia as a probable diamond region on account of its resemblance, in some of its natural features, to Brazil. A few years later, these gems were actually discovered by Humboldt and Rose, on the west side of the Uralian chain, in the gold-bearing alluvium on the banks of the river Adolfskoi, several feet above a stratum containing fossil remains of the mammoth, which has led some eminent scientists to conclude that the diamond of this region was formed since the extermination of this gigantic mammal. Since their first discovery, these precious stones have been found scattered along the western declivity of the Urals, but not in large numbers as in the mines of South America and South Africa.*

There is no country of modern times, unless it is Persia or Brazil, that has a more extensive collection of diamonds, and so many of remarkable size and beauty, than Russia, many of them obtained by conquest, treaty, purchase, or inheritance. The display of wealth in this gem at the London Exposition of 1851 was unsurpassed. Among the exhibits from that country, were a magnificent diadem comprising 1814 brilliants, 1712 rose diamonds, 11 very fine opals, and 67 rubies; besides a bouquet of diamonds made in imitation of the eglantine and the lily of the valley, and a wreath of diamonds representing the bryony bearing pear-shaped emeralds. It is reported that a splendid necklace of twenty-two large brilliants, with pendants composed of fifteen diamonds of large size, forms one of the treasures of the Winter Palace.

The United States. — The gold-producing regions of the United States, extending from Virginia to Alabama, have long been known to yield the diamond, says Mr. Hamlin, but no systematic mining operations have been inaugurated, and the specimens discovered have been the result of accident rather

* A false diamond, which cannot be distinguished by sight from the genuine, is abundant in Siberia, but its use, it is said, is strictly prohibited in Russia.

than design. Itacolumite, a formation in which this gem is found, occurring along this belt, has yielded some good specimens in Alabama, weighing from three to four carats. Gold-miners in the northeastern part of Georgia have occasionally found diamonds in the gravel, and from an examination of one of their mines, this writer is persuaded that the region is a true diamond-field. All the stones from this section of the state are finely crystallized.

Mr. Kunz mentions several localities where these precious stones are found, — in Idaho, Colorado, California, and other gold-bearing regions, but generally of small size and not in sufficient numbers to warrant any extensive mining for them; the garnet districts of Arizona and New Mexico are considered favorable for their production.

Several diamonds of some value have been discovered in North Carolina, while California has yielded this gem in isolated specimens, in nearly twenty different localities, one stone having been known to weigh between seven and eight carats. But the largest native diamond yet known was found in Manchester, Virginia, in 1855. It has been described as octahedral in form, of perfect transparency, and delicate greenish tinge, but with a slight flaw; it weighed before cutting nearly twenty-four carats, and was valued at four thousand dollars. It was cut by Mr. H. D. Morse of Boston, an operation which reduced its weight to eleven and eleven-sixteenths carats, and was considered at one time worth six thousand dollars, but as the color is defective, its commercial value is estimated greatly below that price. Regarded as the largest diamond this country has yielded, it has something above a money value, and should be placed in the National Museum, at Washington.

The Arizona Swindle. — A few years ago, it was reported

that a diamond-field had been discovered in Arizona, and many credulous people were allured to this region with the expectation of making their fortunes by purchasing the right to mine for the diamonds. The report was fabricated and circulated by a band of swindlers, who very shrewdly scattered a quantity of rough diamonds, or stones resembling them, about the designated locality, with the view of entrapping the over-sanguine speculator; but the fraud was detected in season to prevent any general serious disaster.

Thus far the United States has been a large importer of diamonds and other precious stones of the first class, thus proving the wealth and luxury of many of her citizens. A notice in one of the public journals of the seizure of smuggled gems, made at the New York Post-office at one time, will give some idea of the numbers imported. In the list were 132 diamonds, 200 sapphires, and 85 rubies, the whole valued at from ten thousand to thirteen thousand dollars. Though the possession of diamonds by private persons is by no means a rare occurrence, yet foreign writers have absurdly exaggerated ideas of their use in this country. Every one, says Burton, who can afford it wears them, "even hotel waiters and negro minstrels." "The amount imported into the United States," he goes on to say, "has greatly increased since 1849"; this is doubtless correct, "but those introduced are generally of small size, not often more than twelve carats in weight." Streeter says that "diamonds are worn very generally in America by both sexes, in all classes of society and in the streets of large cities, but Americans have scarcely appeared in the market as earnest bidders for large diamonds, and, in spite of the general use of these gems, the great historical specimens are found outside the Republic."

In confirmation of what is correct in these statements just

quoted, as well as for the refutation of what is erroneous, it may be stated that the firm of Tiffany and Company, during the last year (1885), imported a collection of diamonds in the rough, including 882 specimens, weighing in the aggregate about 1775 carats, besides a number enclosed in the diamond-bearing rock, increasing the list to 904 specimens, representing 1876 carats. The collection embraces diamonds of various colors,—pure white, bluish-white, grayish-white, white with yellow spots, light and deep brown, different shades of green, smoky, gray, rose, reddish, yellow, and black. A large part of these stones are of small size, though several specimens exceed ten carats in weight, including a white diamond with striated faces, weighing nearly twelve carats, a smoke-brown exceeding fifty-seven, another brown gem of more than eleven carats, a fine yellow of twenty, a white twin octahedral of more than fifteen, besides several specimens of bort and diamond conglomerates weighing individually from nearly forty-nine carats to more than one hundred and sixteen. A small number of the collection were obtained from Brazil, and the remainder from South Africa.

CHAPTER XII.

HISTORICAL AND REMARKABLE DIAMONDS.

CERTAIN diamonds have played so prominent a part in human affairs that they have become historical, and are entitled to individual names and a special biographical notice. When it is considered that it constitutes the most important gem in regal crowns and other emblems of rank and royal authority, it will be seen there is no material substance which has exercised such an influence upon the destiny of races and nations, or has been connected with so many tragical events, as the diamond. What bloody wars have been waged, what acts of injustice, oppression, and treachery, have been perpetrated, all to gain a diadem, the symbol of sovereign power!

Many others, though with no traditional or historical fame, have an interest on account of their size, beauty, or commercial value, which entitles them to a rank among the celebrated diamonds of the world. There are in all, probably, from eighty to one hundred gems of this species which may be called remarkable for some inherent qualities or acquired fame they possess. These have been described by various writers upon the subject, with a difference of opinion on some points, but with a general agreement of the leading facts in relation to a majority of this list, though in regard to a part of the number there are irreconcilable contradictions respecting their weight, history, and even identity.

Large diamonds have always been rare; it has been estimated that the actual number over thirty carats existing in every part of the world cannot be more than one hundred, of which, it is thought, about fifty are in Europe, and the remainder in Persia, India, Borneo, and, it may be added, on the Western Continent. This number is constantly increasing by new discoveries, and it would be difficult to give an exact estimate at the present time.

A Portuguese dealer of the sixteenth century declared that no specimens of more than thirty-seven and one-half carats ever left India unless by stealth, so strict were the laws to keep all the finest and largest diamonds from being exported. Mawe, writing in 1839, expressed the opinion that the whole number of diamonds in Europe distinguished for their size and beauty scarcely reached half a dozen, and these were in the possession of sovereign princes. Since then the number has been greatly augmented by the discovery of new sources of supply, and the removal of restrictions upon the exportation from their native countries.

Without taking the Braganza into the account, there are only two known diamonds — both uncut — that weigh more than three hundred carats, and seven whose weight exceeds two hundred. There are about twenty with a size exceeding one hundred carats, while all the rest of the catalogue of celebrated diamonds fall below one hundred.

At a sale of the effects of the late Duke of Brunswick, his collection of diamonds included seven ranging from thirty-seven to eighty-one carats in weight. Two of the oldest authentic diamonds in Europe, excepting the Koh-i-noor, — one of twenty-four carats, and the other exceeding that weight, — are said to belong to the Sultan of Turkey.

The Koh-i-noor, or Koh-i-Nûr. — No diamond, probably, has had a more romantic history, or has figured more largely in the affairs of nations and individuals, than the Koh-i-noor, or

"Mound of Light," as the name implies. Tradition assigns it an exceedingly great antiquity, having been found in the Godavery river, Southern India, between four thousand and five thousand years ago, previous to the Indian war celebrated in the great epic, the Mahabharata, and was worn by one of the chiefs who fell in battle on the occasion. Consequently, it is the oldest known diamond in the world.

It came into possession of the family of one of the ancient native princes, the Rajah of Malwar, and was transmitted to his successors through many generations, until it passed into the hands of the Mohammedan conquerors of India, at the beginning of the fourteenth century. It constituted one of the most valuable gems of the imperial treasury of Delhi, until it was carried off by Nadir Shah, the Persian conqueror, in 1739. After the assassination of Nadir, this gem became the property of the Afghan monarchs, and from them was transferred to Runjeet Singh; the Sikh hero of the Punjaub, who had it set in a bracelet, and just before his death, in 1839, he was advised to devote it to Juggernaut, but the act was not consummated, and it was left among his other treasures.

The story is told that Nadir Shah possessed himself of the diamond by artifice. He believed that it was concealed in the turban of the dethroned emperor, since it could not be found in the treasury at Delhi, and on the pretext of restoring the conquered ruler to his dominions, which the wily Persian made the occasion of a grand display, he artfully proposed, as a mark of friendship, to exchange turbans with his imperial guest, an act of courtesy the prisoner did not deem it politic to refuse, and the famous diamond came into the hands of the conqueror, who, on beholding it, exclaimed, "Koh-i-Nûr!"

On the fall of Nadir Shah's extensive empire, Ahmed Shah, the Afghan chief who established a new dynasty, became the fortunate or unfortunate possessor of this ill-omened treasure, "a stone of fate," and from him it descended to his heirs. The last of the line, Shah Soujah, kept this one cherished treasure during his imprisonment and exile until Runjeet Singh compelled him to sell it for one hundred and fifty thousand rupees. After the subjugation of the Sikhs by the English, and the annexation of the Punjaub to British India, in 1849, the civil authorities took possession of the treasury at Lahore, under the stipulation that all the property of the State should be confiscated to the East India Company, and that the Koh-i-noor should be presented to the Queen of England: thus the "talisman of Indian sway passed from the land of its birth to the royal treasury of Windsor Castle."

There is not much doubt, says King, that it is the same diamond mentioned by Baber, the founder of the Mogul Empire of India, in 1526, and the one captured by the Rajah of Malwar, in 1304.* After falling into the hands of Baber it became associated with the most stormy events of modern history. The unfortunate Shah Rokh, when subjected to the most cruel tortures by Aga Mohammed, to compel him to reveal the hiding-place of this coveted treasure, refused to yield it, though suffering the keenest agonies.

Tavernier saw this diamond in the treasury of the Great Mogul, Aurungzeeb, in 1665. Professor Tennant says there is strong probability that it is a part of the original gem of the same name, which was taken from the mines near Raolconda, and was seen by this traveller during his explorations of the Indian diamond-fields. Tennant concurs with the opinion of Professor Nicol that the Great Mogul, the' Koh-i-noor, the

* This author believes its career can be fully authenticated from the time of its possession by the Rajah of Malwar to the present time.

Orloff, and another, a nameless stone, were all parts of one enormous diamond.

The Hindoos have a superstitious belief that this gem brings certain ruin upon the person or dynasty possessing it, and it is a remarkable historical fact that every owner except the last, was the victim of adverse fortune; if a ruler, his own power or that of his line was overthrown, and, adds King, as if its malign influence still accompanies it, not long after the Koh-i-noor became the property of the English crown, the Sepoy mutiny occurred, by which the government came near losing all India.

Tavernier gives the original weight of the Koh-i-noor 793 carats; after it was broken, it was reduced to 279$\frac{9}{16}$ carats, uncut. It is not a little remarkable that there should be any difference in the estimate of the weight after it was cut and recut, yet such is the case. Church places the size at 193 carats Indian cut, and 102$\frac{3}{4}$ London cut; Westropp gives the weight 186$\frac{1}{4}$ and 103$\frac{3}{4}$; King, 186$\frac{1}{2}$ and 102$\frac{1}{2}$; Emanuel, 186 and 106$\frac{1}{16}$ carats. When first imported into England it was the largest in Europe except the Orloff, but at the present time it is surpassed by several others.

The recutting of this diamond was performed in London by steam power, under the direction of artists from Amsterdam, and occupied thirty-eight days, at a cost of forty thousand dollars. It is cut in the form of a brilliant, which involved great waste of material; and, though the beauty of the gem may have been enhanced by this operation, which is a matter of doubt, and some of its defects removed or rendered less apparent, yet, in the opinion of many, its value as an historical monument has been greatly diminished.

The Great Mogul. — This is the largest authentic diamond ever yet discovered; the only one surpassing it, denominated the Braganza, or the King of Portugal's, is supposed to be a topaz. It was discovered in the mines of Gani, or Coulour, in 1650, and came into the possession of the vizier of the King of Golconda, who had amassed great wealth by farming the diamonds. Becoming the object of jealousy to his royal master, the minister escaped from court, taking with him all his treasures, and found his way to the capital of Aurungzeeb, one of the most celebrated of the Mogul rulers. Here the fugitive vizier became a favorite by his munificent presents to the emperor, including this famous diamond, which was from its imperial owner, called the "Great Mogul."

Its weight at that time, according to Tavernier, was seven hundred eighty-seven and one-half carats, but as it was badly flawed it was decided to have it cut — an operation performed by an unskilful Venetian lapidary named Borgia, or Borghis, who reduced its weight nearly one-third. It was cut in the form of a half egg, and covered with facets like a rose diamond. The emperor was so offended at the great waste of the stone, that he not only refused to pay the operator his stipulated price, but imposed a fine of ten thousand rupees for damages. The unfortunate Venetian, who evidently understood neither his business nor the temper of despotic princes, was glad to escape with his life. The same or a similar incident has been related of the Koh-i-noor, and Dieulafait thinks it was this diamond and not the Great Mogul which Tavernier saw at the emperor's court.

The two gems have sometimes been confounded, and the traditions of the one have been merged with those of the other; yet they were perfectly distinct, having little in common except their great size and their presence in the imperial treasury.

The weight of this diamond, like the Koh-i-noor, has been differently estimated at 900, 793⅓, and 787½ carats, before cutting; $279\frac{7}{16}$, 297, and 280 carats, after cutting. It was of the first water, and valued at one million six hundred and eighty thousand dollars.

This diamond, like its supposed twin brother, is connected in history with some of the wars and revolutions in India, and, during the Persian invasion, was lost sight of, which afforded some ground for the general opinion that it now belongs to the Persian crown, but King says there was no stone of this size and pattern among the drawings of the Shah's diamonds recently brought to England, and it is more likely the Great Mogul has been lost, and if so, it may come to light by some unexpected event, unless it has been cut into smaller stones to elude detection.

Tavernier entertains his readers with an account of the ceremony of exhibiting the crown jewels at the Mogul's court, in which he says they were brought into the imperial presence by the custodian on lacquered trays, covered with brocade, when they were counted three times, and a list was made out by three scribes before they were returned to their cabinets.

The King of Portugal's, or the *Braganza*, called also the Brazilian. — If this is a genuine diamond, of which there are great doubts, it is the largest on record. Its weight in the rough (it has never been cut), has been estimated both at 1680 and 1880 carats. Probably the former weight is nearer the truth. Cutting would doubtless reduce its size two-thirds, leaving then a stone of five hundred and sixty carats, twice as large as the Great Mogul. It has been rated at a fabulous price, even for a true diamond, varying from about thirty million to nearly sixty million pounds. There is no infallible standard by which to estimate the worth of an uncut stone, therefore all such valuations are mere guesses.

This geological marvel has given rise to a difference of

opinion as to its true character, some judges believing it to be a topaz, others rock-crystal; but as the government declines having it tested, the doubt cannot be cleared up. It is about the size of a hen's egg, and of a deep yellow color. It was found in Brazil in 1741, by a convict or a slave..

Some writers have considered the Braganza a different stone from the King of Portugal's, and identical with the Portuguese Regent, a gem of two hundred and fifteen carats. King mentions two uncut diamonds of about this weight belonging to the crown, and Streeter says the Regent and the Braganza, which have been regarded identical, are different stones.

The Nizam. — The history of this diamond is involved in the obscurity of a doubtful tradition, and though one of the largest on record, there is but little positive knowledge concerning this gem, except its size, and present owner, the Nizam of Hyderabad, a semi-independent ruler of the Deccan, for whom it was named. It was discovered in the so-called Golconda mines, which are in the dominions of this prince, and weighs, in the rough, three hundred and forty carats, after having been broken by some accident, during the great Indian revolt; before this casualty, its weight is supposed to have been four hundred and forty carats.

The Rajah of Matan. — This diamond was found at Landak, on the western coast of the island of Borneo, during the last century. It is said to be egg-shaped, of the purest water, and weighs either three hundred and sixty-seven or three hundred and eighty-seven carats, according to different estimates; cutting would probably reduce it below the Orloff. As its true character has never been fully established, its large size awakens suspicion of its genuineness, but since the owner is the Rajah of Matan, there is little probability that it will ever be proved beyond a doubt. It has remained in the ruling family during four generations, and is regarded as a kind of tutelary deity, with miraculous healing powers. The Dutch government negotiated for its purchase, offering the rajah two gunboats, with all their stores and equipments, and two hundred and fifty thousand dollars in money; but the owner could not be induced to part with it on any terms, from the belief that the perpetuity and success of his line depended upon his retaining possession of this gem. It is said to have been the cause of a destructive war, during some period of its history.

The Orloff. — This celebrated gem has its tale of romance, intrigue, and crime, its remarkable history and conflicting biographies, which place it alongside of the Koh-i-noor for dramatic interest, while for size and beauty it holds a first rank among European diamonds.

The history of the Orloff has so many different versions, it is extremely difficult, if not impossible, to decide which is authentic, especially as its career has frequently been merged in that of the Great Mogul, and also in a Persian diamond, called the "Moon of the Mountains." There is no doubt that it is an Indian gem, that it found its way to Holland, thence to Russia, where it was placed in the imperial sceptre, and is considered the most remarkable diamond in the Russian regalia, and one of the most famous in Europe; but in relation to its romantic history before it became the property of the crown, the records diverge. It was purchased at Amsterdam, by Prince Orloff, whose name it bears, for Catherine II., according to one historian, but, following the annals of some other writers, it has a more devious and adventurous career, having once constituted one of the eyes of a famous Indian idol, from which it was plundered by a French deserter, and sold to an English sea-captain; from him it passed to a London Jew, then to a Greek, who offered it for sale to the Empress of Russia, but, Catherine declining to purchase the diamond, it was bought by Prince Orloff, for four hundred and fifty thousand dollars, and a life annuity worth twenty thousand dollars, and presented as a gift to his sovereign. The patent of nobility said to have been given to the Greek merchant was conferred upon the owner of the "Moon of the Mountains," with whose history the Orloff has been confounded. What became of the companion eye of the Hindoo god has not been ascertained. Streeter discredits the story of the idol, and says the Orloff was brought from Seringapatam a fortified island in Mysore, and that its true name was Koh-i-Tûr. Another conjecture about its early antecedents is that it formed one of the trophies brought away from Delhi by Nadir Shah, but was lost sight of for a time during the fierce struggles which ensued on the death of Nadir, was recovered and sold to an Armenian by one of the Afghan generals, and through the Armenian it reached the Russian treasury; its course up to that event was marked by deception, theft, and murder. It is said the original name, given by the Persians, was Koh-i-Tûr, which has been rendered "Mount Sinai," and that, with the Koh-i-Nûr, it constituted the eyes of the bird in the famous Peacock Throne.

The Orloff is about the size of a pigeon's egg, of a yellowish tint, with a weight of from one hundred and ninety-three carats to nearly one hundred and ninety-five. The identity of this diamond with the Great Mogul is disproved by the great difference in size.

The Austrian Yellow, called also the *Grand Duke of Tuscany's*, and *Florentine Brilliant.* — The identity and history of this precious stone have been the foundation of much controversy, which has not yet been settled to the satisfaction of all parties.

It has its tale of woe if, as is sometimes said, it once belonged to Charles the Bold and was lost, together with his other jewels, on the disastrous field of Grandson. It was found by a Swiss soldier, who, after the conflict, ignorant of its real value, sold it to a priest for a florin, and from him it passed to a Genoese merchant, who disposed of it to the Grand Duke of Tuscany. Subsequently, the gem came into the possession of Pope Julius II., one of the Medici family, and by him was presented to the Austrian crown in the reign of Maria Theresa. That the stone is cut after the Indian fashion, is produced as counter evidence that it came into Italy from the East, therefore never belonged to the Duke of Burgundy, and could not have been found on the field of Grandson. Tavernier saw this diamond in Florence in the middle of the seventeenth century, and from this date its authentic history begins.

The Austrian is of a decided yellow tint and ranks next to the Orloff for large cut diamonds found in Europe; it is a double rose, faceted all over, and weighs one hundred thirty-nine and one-half carats; different estimates represent its value from two hundred thousand dollars to nearly eight hundred thousand.

The Great Sancy. — This diamond has been very appropriately called the sphinx among diamonds; indeed, its history is involved in bewildering contradictions, which it is impossible to unravel with only the glimmering light which illumines its record. It appears and disappears in the most mysterious manner for a period of four centuries, and has been the subject of more conflicting statements than any other or perhaps all other diamonds known to literature or science. There is no way of reconciling these contradictions but by admitting there was more than one diamond called Sancy.

Tradition refers this gem to Charles the Bold, its first

European owner, which, with the Austrian Yellow, was lost on the day of his memorable defeat; and here begins the complicated knot of difficulties never yet completely disentangled.

The diamond was named for Nicholas de Harlay, Seigneur de Sancy, at one time ambassador of Henry IV. of France to the court of Queen Elizabeth. The Baron de Sancy, it is said was the owner of two large diamonds, one having been purchased when he was ambassador to the Ottoman court, and the other was taken in pledge from Don Antonio, the pretender to the Portuguese crown, for a loan of one hundred thousand livres, which, however, was never redeemed; this accounts for a part of the confusion pertaining to the subject.

De Sancy, as the story goes, in order to raise funds for Henry to enable him to prosecute his wars, pledged the famous diamond known as the Sancy to the Jews of Metz, and sent it thither by a trusty messenger, with the understanding that if the latter were attacked by banditti, a danger imminent in that turbulent period, he should swallow the gem. The catastrophe did occur, the messenger was murdered, but his master, having confidence in the fidelity of his servant, recovered the body, and found the diamond in the stomach. A little different version makes the king send the gem to Harlay, who was then in Switzerland, but for the same purpose.

A different account, which probably applies to another diamond bearing the same name, states that Seigneur Harlay sold the gem when minister to England, and that it is mentioned in the inventory of the crown jewels in the Tower of London, made in 1605. It remained in the possession of the royal family until 1669, and is mentioned by Henrietta Maria. This Sancy was the one sold by James II., during his exile, to Louis XIV., for one hundred and twenty-five thousand dollars; it formed a part of the crown jewels at the time of the inventory of 1791, when its value was fixed at two hundred thousand dollars. It was stolen at the robbery of the Garde Meuble, and all traces of the diamond were lost until 1830, when it reappeared as the property of a merchant, but by what means he obtained possession of it, or where it had been concealed, is involved in mystery.

A Sancy diamond became the property of the Demidoff family of Russia, and was the subject of a lawsuit, in 1832, with the director of mines in Switzerland. Pending the trial, there was the most conflicting testimony concerning its history, which could be accounted for only by admitting there had been more than one diamond of this name in the Harlay family. The Demidoff gem was sold, in 1865, to an Indian millionnaire, or Parsee merchant, of Bombay. But the wanderings of this jewel did not end here; it was sent to Paris, where it was exhibited in 1867, then returned to India, when it was purchased by a native prince, together with many of the jewels of the Empress Eugénie, and was worn on the reception of the Prince of Wales during his visit to the East. The last possessor of this mysterious gem has since died, an event which may again change the fortunes of the Sancy diamond, and start it on its devious travels. Another account varies, in some points, with that just related, and coincides with others; as, that it was sold at Lucerne in 1492, came into the possession of Portugal in 1594, and was sold to De Sancy; but this leaves one hundred years of its history unaccounted for, which may include its career at the Burgundian and English courts. After remaining in the Harlay family more than a century, it was sold to the Duke of Orleans, Regent of France. It disappeared during the French Revolution, but was recovered and bought by Napoleon I., who sold it to Prince Demidoff.

Since that date, the different narratives are nearly identical, and probably authentic.

Mr. Hamlin favors the opinion that there have been three different diamonds bearing the name of Sancy: first, the one belonging to Charles the Bold; second, the one sent to the Jews at Metz, but never redeemed, and nothing further is known of its history; third, the Demidoff gem, sold to an Indian purchaser. The weight of the great Sancy has been given as fifty-three and one-half, and also fifty-four carats, Indian cut.

The Little Sancy.—This gem, worn by the bride on her marriage to Prince Albert of Prussia, led to a mistake in the public journals, where it was reported as the famous Sancy diamond. It is a brilliant of thirty-four carats, and was purchased by Prince Frederick Henry of Orange, grandfather of King Frederick I., in 1647, and thus passed into the royal treasury of Prussia. It is believed to be one of the diamonds owned by De Sancy, bearing his name, but distinct from the "Great Sancy."

The Diamond of Charles the Bold.—The jewels of this prince, who was an indefatigable collector, have been the source of much bewildering speculation on the subject of historical gems, which appears no nearer solution after the lapse of centuries than at first. These treasures, lost on the battlefield, have become identified with the fortunes of their owner, hence they are invested with a romantic interest beyond many others, and have acquired a celebrity in the literature of precious stones. The history of this diamond seems to be merged in that of the Great Sancy, at least through a part of its career. It is said to have been found by a Swiss soldier, who sold it for a florin. Afterwards it was disposed of to the Bernese government for three francs, and passed, with other valuable trophies of the victory, to Fugger.

This diamond was considered one of the largest in Christendom, and was described as a pyramid, five-eighths of an inch square at the base, and with the apex cut into a star, in relief, presenting four rays. It was recut by Berquem and set with three large balas-rubies, styled "The Three Brothers." Four baroque pearls, weighing half an ounce each, formed the pendant. The jewel was sold to Henry VIII. of England, and was given by Queen Mary to Philip II., a descendant of the House of Burgundy. The remarkable part of the story is that a diamond mounted in this ornate style should have been twice sold for a trifle.

The Pitt or Regent. — This diamond, before cutting, ranked next to the Great Mogul in size, weighing four hundred and ten carats. It was found in the Indian mines at Puteal, in 1701, by a slave, who concealed the discovery, and fled with it to the coast, but only to meet with a tragical end; for, being decoyed on board an English vessel, he was robbed and thrown overboard. The captain who committed the double crime sold the diamond for the paltry sum of five thousand dollars, spent his dishonest gains in dissipation, and closed his career by hanging himself.

The next we hear of this prize is that it was bought of a Parsee merchant by Thomas Pitt, governor of Fort George, Madras, and grandfather of the celebrated William Pitt, for the sum of ten thousand dollars, or sixty-two thousand five hundred dollars, according to widely different statements. This purchase was the occasion of much scandal and satire in England at the time, reflecting — unjustly it is thought — upon the integrity of the governor. Pope alludes to the transaction in the "Man of Ross" in the lines, —

> "Asleep and naked as the Indian lay,
> An honest factor stole the gem away."

The diamond was cut in London by hand, as a brilliant, weighing from one hundred and thirty-six to nearly one hundred and forty carats, as variously estimated, at a cost of twenty-five thousand dollars. The value of the fragments has been reckoned from seventeen thousand to forty thousand dollars. The work required two years, while the operation of cutting the Koh-i-noor was performed in thirty-eight days; the difference in time being due largely to the difference in the agencies employed, — for the former, manual labor; for the latter, steam-power.

As the diamond was not purchased by the English crown, it was sold to the Regent of France in 1717 — whence the name Regent — for six hundred and seventy-five thousand dollars, a price considered much below the true value; it was estimated at the inventory made by the decree of the Constitutional Assembly at two million four hundred thousand dollars. At the robbery of the Garde Meuble, the Regent was stolen, but recovered by a communication from one of the party of thieves revealing the spot where it was concealed; the reason given for its surrender was that the gem was so well known it would not be safe to offer it for sale.

The Regent, it has been said, laid the foundation of Napoleon's brilliant career, since by pledging it to the Dutch government, he obtained funds for prosecuting his military operations and for the establishment of his power; it was subsequently redeemed, and, after he became emperor, adorned his sword of state. It is a remarkable coincidence in the history of this diamond that it should have been connected with the fortunes of two eminent contemporary men who were bitter foes, — Pitt and Napoleon; for there is not much doubt that this gem had an influence in establishing the prosperity of the Pitt family — consequently, the success of the great English statesman. No mention is made of the Regent in the inventory made by Napoleon in 1810, nor in any subsequent official report of the crown jewels, though it was exhibited at the Exposition of 1855, and is claimed to be the most conspicuous gem in the French crown. This celebrated diamond is pre-eminent for its symmetrical form, its transparency, purity, and beauty.

The Hope Blue. — One of the most common occurrences connected with the history of precious stones is to find writers on the subject disagreeing in their statements; therefore, we have another illustration in the gem known as the Hope Blue, of the inexplicable confusion of ideas regarding the identity of celebrated diamonds. Three different blue stones of this species have been mentioned as belonging to France,— the French Blue, the Tavernier Blue, and the Hope Blue; and the difficulty presents itself whether these are identical, or whether there were three or even two blue diamonds, or whether the genuine Hope is really in existence or has been lost. Streeter and some other writers believe there was one large blue diamond cut into three stones, and that the Hope was one of them. The Hope Blue, of 44¼ carats, exhibited at the London Exposition of 1851, among the French jewels, is described by Professor Tennant, who says it combines the beautiful blue of the sapphire with the prismatic "fire" and brilliancy of the diamond. It was mounted as a medallion, with a border, *en arabesque*, of small rose diamonds surrounded by twenty brilliants, all of the same size and of the first water. This famous gem, unique of its kind, the most beautiful specimen of blue diamond known, was purchased by Mr. Hope, an English banker, for the reputed sum of sixty-five thousand dollars; a price, it is thought, far below its real value. The only other gem of the kind which can approach it in beauty is the *Blue*

Diamond, weighing thirty-six carats, seen in the Munich collection.

The Tavernier Blue was brought from the Coulour mines, with twenty-five other large diamonds, by the traveller whose name it bears, and sold to Louis XIV., for five hundred thousand francs and a patent of nobility. Its original weight was sixty-seven and one-half carats, reduced by the lapidary to forty-four and one-fourth.

The French Blue has been represented as a fine gem belonging to the French crown, with a weight and value equal to that of the Tavernier; it is supposed it was split, and one of the pieces formed the Hope. These accounts so far coincide as to leave but little doubt of the identity of the Tavernier Blue and the French Blue, which was lost, it is believed, at the robbery of the Garde Meuble. The most remarkable feature about these gems is that there should exist three blue diamonds of precisely the same weight.

One of the diamonds of the late Duke of Brunswick claims affinity with the Hope Blue, on account of its remarkably brilliant tint, and another blue specimen, forming, says Mr. Streeter, one of the rarest jewels in the world, seen in the fashionable circles of London, is referred to the same origin. It is possible that all these blue diamonds, if they ever existed, once formed a part of an immense stone which, by some unknown agent, was separated into several smaller ones.

The Great Table.—The diamond seen by Tavernier in India called by this name, weighed two hundred forty-two and one-half carats, and presented the appearance of having been split, which suggested the idea that the *Russian Table*, of sixty-eight carats, may have been a part of it. No European expert has ever seen it since Tavernier's day, and even its present owner is not known; it may have been one of the "Three Tables" mentioned by this traveller carried off by Nadir Shah, and may turn up either in Persia or Afghanistan.

The Nassack.—This trophy was gained by the English during their war with the Mahrattas, and is said to have been taken from one of the famous cave temples near Nassack, on the Upper Godavery, to which it had been presented as a sacred offering by the natives. The diamond was delivered to the Marquis of Hastings, then Governor of India, who presented it to the East India Company. It was sent to England, and sold to Messrs. Rundell and Bridge, of London, for a sum far below its real value, and finally came into the possession of the Duke of Westminster, who placed it in the hilt of his sword. Its weight was reduced by re-cutting from eighty-nine and three-fourths to seventy-eight and three-eighths carats, but it gained vastly in brilliancy; after this operation it was valued at a price from forty thousand to fifty thousand dollars.

The Ascott brilliants, or diamond ear-rings presented to Queen Charlotte by the Nabob of Ascott, were bought by the Duke of Westminster at the same time.

The Pigott.—This was a brilliant of the purest water and one of the finest in Europe; its weight has been variously estimated, but the preponderance of testimony favors eighty-two and one-fourth carats. It was an Indian gem, introduced into England by Lord Pigott, Governor of Madras, about the year 1775, and sold at the beginning of the present century to Rundell and Bridge, who estimated its value at one hundred and fifty thousand dollars. There is a little romance connected with this beautiful gem. It was purchased by Ali Pasha, who cherished it with remarkable devotion, and, when mortally wounded, gave orders to have this diamond destroyed, and his favorite wife strangled, to prevent their falling into the hands of his enemies. The order respecting the diamond was executed in his presence, but fortunately the queen of the harem escaped the bloody decree respecting her fate.

The Shah. — This diamond formed a part of the Persian regalia from remote times, until it passed to the Russian treasury, where it now belongs. No authentic record of its earliest history has been transmitted, but there is little doubt that it originated in India, whence a large number of the celebrated diamonds came. It was one of the valuable gems plundered from the treasury of Nadir Shah after his death, but it was recovered and presented to the Emperor Nicholas by Prince Kosroes, son of Abbas Mirza, when on a visit to St. Petersburg, in 1843. The diamond is a long prism, of the first water, and weighs eighty-six carats after atting. It is engraved in Aralo-Persian characters, with the names of Akbah Shah, Nizam Shah, and Ali Shah, three Persian rulers, "Lords of Irostan." It is said that this diamond and the Akbah Shah are the only ones brought from the East that have been engraved.

The Akbah Shah. — This gem was found among the treasures of the Mogul emperors, and was engraved on both sides in Arabic; on one was the inscription, "Shah Akbah, the Shah of the World, 1028 A. H.," corresponding to 1650 A. D.; on the other side was engraved, "To the Lord of Two Worlds, 1039 A. H." — 1661 A. D. The diamond was lost sight of at the close of the seventeenth century, but has recently come to light, so it is stated.

The Shah Jehan. — There are strong probabilities that this diamond is identical with the Akbah Shah. It came into public notice in Turkey, a few years ago, where it was called the "Shepherd's Stone," and was purchased by an English gentleman, who had it recut in London. It was an engraved stone, weighing one hundred and sixteen carats, but the cutting entirely effaced the inscription, and reduced the size to seventy-two carats. As it bore an inscription, it must have been either the Shah, or the Akbah Shah, since they were the only engraved oriental diamonds known; it could not have been the Shah, and as it disappeared for a time, and has recently been found, there seems to be but little doubt that the same diamond appears under two names. The Shah Jehan was sold to the Prince of Baroda, of India, who became the possessor of several notable gems of this kind.

The Darya-i-Nûr. — This Persian name, signifying, "Sea of Light," expresses its remarkable lustre, for which the diamond is distinguished; it is rose-cut and weighs one hundred and eighty-six carats. It was captured at Delhi by Nadir Shah, and, on account of its great size and brilliancy, constitutes one of the finest gems in the large collection of the Persian treasury.

The Tay-e-Mah. — "Crown of the Moon" — is a diamond but little inferior to the "Sea of Light" in size and splendor; it weighs one hundred and forty-six carats. Some authors have thought it might be the Great Mogul but it differs from the Mogul in many essential points. Opinions are not uniform in regard to its early history; according to one view, it was found in the Indian mines on the river Mahanuddy, in 1809, and according to another, it was carried off in 1739, by Nadir Shah,— the freebooter of the East, upon whose shoulders rests the crime of stealing nearly all the famous jewels of his times,— thus making it appear that it was captured before it was discovered,— a remarkable feat even for Nadir. Another theorist tells us that this gem was owned by Shah Rokh, who was tortured by Aga Mohammed to compel him to surrender it, a method by which many other valuable gems were added to the Persian crown.*

Sir John Malcolm, who, when minister to Persia, inspected the regalia, says the Darya-i-Nûr was considered to have the finest lustre of any diamond known, and, with the Tay-e-Mah, constituted the most illustrious ornament of the Persian regalia. Some of the other magnificent diamonds mentioned by this official, besides the "Sea of Light," and the "Crown of the Moon," set in a magnificent pair of bracelets, valued at five million dollars, are the "Sea of Glory," sixty-six carats; the "Mountain of Splendor," one hundred and thirty-five carats; the "Throne," and the "Sun of the Sea," which once ornamented the Peacock Throne at Delhi.

The "Moon of the Mountains." — This diamond was torn from the plumage of the "Peacock," and carried off to Persia, and its subsequent history is mixed up with the blackest crimes on record.

After the assassination of Nadir Shah, and the plunder of his treasures, an Afghan soldier fled with this gem to Shat-el-Arab, on the Red Sea, the emporium of trade between the East and the West, for the purpose of disposing of this precious stone with others robbed from the treasury of Nadir, including an emerald of rare beauty, a fine ruby, a magnificent sapphire called the "Eye of Allah," and other valuable gems. An Armenian trader, named Shaffras, opened negotiations for their purchase; but before the bargain was completed, the Afghan, becoming alarmed lest his robbery should be found out, fled to Bagdad, where he disposed of his treasures to a Jew, for the trifle of two thousand five hundred dollars.

Shaffras followed the soldier to Bagdad with the view of securing the gems, and, learning they were sold, he murdered

* A similar incident is related of the Koh-i-noor.

the Jew and carried off the jewels. But his crimes did not stop here; the Afghan was first despatched for fear he might reveal the assassination of the Jew, and then the two brothers of the murderer, who were his partners in trade, that he might reap all the profits arising from the sale. After the commission of these horrible deeds, Shaffras found his way to Holland, where he established himself as a dealer in precious stones. A report of the valuable diamond in his possession reached the Empress Catherine II., who invited the owner to St. Petersburg, with the object of arranging a bargain for its sale. This invitation was accepted, but before negotiations were completed, the Armenian suddenly disappeared with the gem, probably from fear of detection, and was not heard from until several years afterwards, when he was found in Astrachan. Efforts to secure the diamond were made by the Russian government, which eventually resulted in its purchase for an immense sum; Shaffras was himself afterwards murdered by his son-in-law, a just retribution for his crimes. This tragical story has been connected with another diamond, the "Shah"; but it has very slight grounds on which to rest, since this gem passed directly from the Persian crown to the Russian.

The Abbas Mirza, sometimes called *Jehungheer Shah.*—This diamond, whose weight is said to be one hundred and thirty-eight carats, is supposed to belong to the crown jewels of Persia. It has no record until the capture of Cûcha, in Khorassan, by Abbas Mirza in 1832, and it attracted no public notice, says Streeter, until the meeting of the British Association in 1851. It has been thought it might be a part of the Koh-i-noor, or the Great Mogul; its weight is one hundred and thirty-eight carats.

The Ahmedabad.—Nothing is known of this diamond except that it was brought from the East by Tavernier, and weighed one hundred and fifty-seven and one-fourth carats, reduced to ninety-four and one-fourth by cutting. It is supposed to belong to the imperial treasury of Russia, but if this is true, it is remarkable that a diamond of that size should not be known outside the Empire.

The Turkey, I and II. — These gems, represented as weighing one hundred and forty-seven, and eighty-four carats, respectively, belong to the Turkish regalia, but little or nothing more is known of them.

The Polar Star, formerly owned by Joseph Bonaparte and purchased by Paul III. of Russia, is a brilliant of forty carats, and is distinguished for its superior lustre and perfect purity.

The Pasha of Egypt, bought by Ibrahim Pasha, for one hundred and forty thousand dollars, is a brilliant of the same size as the Polar Star, and has been considered the finest diamond in the Egyptian collection.

The Coulour. — But little is known of some of the historical diamonds except their names and possibly their size; hardly anything of their origin or present owners. Such was the Coulour, a gem weighing fifty carats, brought to Europe by Tavernier from the mines of India.

The Pear, of about the same size, seen by this traveller among the jewels of the Great Mogul, was reckoned one of the trophies of Nadir Shah, and has been lost sight of in the vast Persian collection, or has been recut and cannot now be identified.

The Tavernier, A, B, C, three diamonds sold to Louis XIV., were supposed to be lost at the robbery of the Garde Meuble, though there is some probability that the beautiful stone, weighing fifty-one carats, purchased by Napoleon III. for the empress, in 1860, may have been one of them, recut as a brilliant.

The Crown, a diamond of the size of thirty-two carats, valued at sixty thousand dollars in the inventory of 1791, was placed in the Golden Fleece of the French regalia.

The Savoy, bequeathed to the crown of Savoy, by Queen Christina, in 1662, was a table diamond, weighing fifty-four carats, and set in antique style, with large pearls..

The Eugénie. — As the size of this diamond corresponds to that of one of the Taverniers, it has been thought by some connoisseurs to be identical with it, though this circumstance alone would not be regarded as very strong evidence. This gem has acquired a romantic celebrity from its connection with the fortunes of royal personages. It was once used as an ornament for the hair by Catherine II. of Russia, who presented it to one of her subjects, as a reward for his distinguished public services; subsequently, the Russian gem came into the possession of the French Emperor, and was worn by the empress as the centre brilliant in a diamond necklace. After the fall of Napoleon III., the Eugénie was sold to the Prince of Baroda, for seventy-five thousand dollars; but, as if some malign influence passed from this treasure to its possessors, this Indian prince was afterwards deposed, and the diamond has disappeared.

The Dresden Green. — Nothing is known of the antecedents of this exceedingly rare and beautiful diamond, but it is thought to be of Indian origin; it constitutes one of the most conspicuous gems in the fine collection of the Green Vaults, at Dresden, and is distinct from the "Green Brilliant," which originally belonged to the Elector Augustus of Saxony, and was worn as a button for a hat-band. The Dresden Green weighs, according to Grasse, the director of the vaults, forty and one-half carats,— King says thirty and one-fourth,— and is valued at one hundred and fifty thousand dollars.

Two large brilliants, of nearly forty-nine and thirty-nine carats, respectively, are included in this magnificent collection, the larger being known as the *Dresden White*, or the Saxon white brilliant, which is classed among the finest in Europe for perfection of form and superiority of lustre. The *Dresden Yellow* diamonds are four brilliants of great beauty, each weighing about thirty carats.

The English Dresden, taking the name of its owner, Mr. E. Dresden, was discovered in Brazil, in the same region which yielded the "Star of the South," and is supposed to have formed a part of another diamond. This stone, weighing, in its natural state, one hundred and nineteen and one-half carats, was reduced by the lapidary to seventy-six and one-half carats, and is described as of a drop-shape and absolutely faultless, a distinction seldom known to belong to diamonds, while its purity and lustre are so extraordinary that when placed beside the Koh-i-noor, the latter appears of a yellowish tint, and inferior in brilliancy. This magnificent jewel, failing to obtain a purchaser in any European market, found its way to India, and was bought by an English merchant of Bombay. Its subsequent history, says Streeter, is, in a certain way, mixed up with the American Rebellion. The purchaser was a dealer in cotton, and this commodity, advancing in price in consequence of the American war, brought him a large fortune, a part of which he invested in the diamond trade. Among other specimens, he purchased the English Dresden, for a great price, but, in consequence of the unexpected close of hostilities, resulting in the fall of the price of cotton, the Bombay merchant was ruined financially, and, soon after dying of grief, his coveted treasure passed into the hands of the Prince of Baroda.

It is a remarkable coincidence, or a series of coincidences, says this writer, that the Star of the South, and the English Dresden were both found near the same place in Brazil, about the same time; both were sent to Amsterdam to be cut by the same jewellers; both were forwarded to the same agency in India, and both were purchased by the same prince.

The Cumberland, of thirty-two carats, was bought by the City of London, for fifty thousand dollars, and presented to the Duke of Cumberland, after his victory at the battle of Culloden, in 1746. It is said to have been returned to the crown of Holland, by the Queen of Great Britain, in consequence of claims advanced by the House of Hanover.

Star of the South.—This is a Brazilian gem, discovered in 1853, and thought by some to have belonged to a group of diamonds, a supposition which has been questioned from the fact that these crystals are usually found isolated, though occasionally they occur in geodes. This diamond exhibits a rose color of a slightly yellowish tint, with a remarkable play of prismatic hues, and ranks among the largest extant; the original weight, two hundred and fifty-four and one-half carats, was reduced to one hundred and twenty-five by cutting. It was bought in Paris and exhibited in London in 1862, and at the French Exposition of 1867, where it attracted general attention. After travelling to India and back to Paris, it was finally returned to the East, and was purchased by the Mahratta ruler of Baroda, for four hundred thousand dollars. Emanuel says it belongs to Mr. Costar, of Amsterdam, while Burton has assigned it to the Pasha of Egypt.

The Chapada, another Brazilian diamond, discovered in 1851, weighs eighty-seven ond one-half carats.

The Star of Sarawak, of sixty-eight carats, and the first water, and the *Bantam*, thirty-six carats, were both found in Borneo; the latter, owned by the King of Bantam, and priced at more than fifty thousand dollars, is thought to belong to the Dutch regalia.

The Bazu, owned by a Dutchman of that name, a gem of one hundred and four carats, was obtained from the Coulour mines, India, and brought to Europe by Tavernier. The interior of the stone contained eight carats of what appeared to be decayed vegetable matter, a peculiarity which might be regarded an indication of the origin of the diamond. A similar specimen is seen in the British Museum.

The Raolconda, an Indian diamond mentioned by Tavernier, was nearly of the same size as the Bazu.

The Hastings. — This gem was the occasion of a good deal of scandal pending the trial of Warren Hastings, and gave rise to a street ballad of the times, in which George III. and the governor-general were travestied with great freedom. The diamond was given to the king, not by Hastings, as represented, but by the Nizam of the Deccan, in 1786. It has been described as a fine specimen, but it cannot now be identified in the collection of the crown jewels.

The Stewart. — This diamond had no rival among the South African diamonds until the discovery of the Porter Rhodes, in 1880, and was surpassed in size only by the Great Mogul, the Matan, and the Nizam. It weighs, in its undressed form, two hundred and eighty-eight and three-eighths carats; the tint is a pale yellow.

The Porter Rhoades, named for its owner, is a bluish white gem found at the Kimberly mines, South Africa. Its weight in the rough has been said to exceed that of the Stewart, which, if this estimate is correct, must have been near three hundred carats. Both Streeter and Church give its size, after cutting, undoubtedly, one hundred and fifty carats.

The Heart formed the centre of a rose composed of twelve

large diamonds and an equal number of pearls, for a jewel in the turban of Baber, the first Mogul emperor.

The Star of Diamonds. — This brilliant title was won by the resplendent appearance of the gem under the microscope, when it exhibited a view resembling crests of mountains illuminated by the most vivid colors of the rainbow. It is an African stone, of one hundred and seven and one-half carats.

The Napoleon. — But little is known of this gem ; it is said to have been worn by the emperor in the hilt of his sword at his marriage with Josephine, though there is no mention of it in the inventory of the crown jewels made in 1810.

A diamond brought from the East Indies by Hon. William Hornby, Governor of Bombay, and called the *Hornby Diamond*, is supposed to belong to the Shah of Persia. The gem known as the *Antwerp Diamond*, was sold in 1559, to Philip II. of Spain, for eighty thousand crowns. One of the largest diamonds found in Brazil was called the *Patrochino.*

The prolific mines of South Africa have yielded some specimens of large size, and may, possibly, afford others which will surpass any on record, but the greatest number of celebrated diamonds has, hitherto, been found in India. In a paragraph of one of the public journals, it was stated that a diamond of immense size had been recently shipped from South Africa to England, which would afford a gem weighing two hundred carats cut as a brilliant, or about three hundred cut lozenge-shape.

In addition to those previously named from the South African mines are *The Tennant*, so called for the late scientist of this name, a diamond of a yellowish tint, weighing sixty-six carats, and forming a brilliant ornament in the English regalia ; the *Jagersfontein*, an uncut gem, of two hundred and nine and one-fourth carats ; *Du Toit I.*, weighing two hundred and forty-four carats; *Du Toit II.*, representing one hundred and twenty-four; the *Star of South Africa*, a stone of forty-six and one-half carats, besides several others of superior weight, including the *Tiffany Diamond No. I.*, owned by the firm of that name, a gem of great beauty, which, cut as a brilliant, weighs one hundred and twenty-five and three-eighths carats, and is valued at one hundred thousand dollars.

An article published in Science, May, 1884, written by Mr. G. F. Kunz, describes certain remarkable round diamonds from Brazil, the largest weighing forty-one and three-fourths carats, which were exhibited at the Amsterdam Exposition, and subsequently purchased by Messrs. Tiffany and Company. One of them was bought by Krom Mun Nares Varariddhi, Prince of Siam, during his late visit to this country.

Two other diamonds in the possession of this firm are interesting on account of their peculiar characteristics; one, weighing six and three thirty-seconds carats, has eighteen facets, of which four of the top and the table are white, and four are decidedly black, and four on the back are white, while the remainder and the culet are black. In its native state, the diamond was a jet-black, but when submitted to the lapidary's operations, the interior of the crystal was found to be perfectly white with the exception of an occasional inclusion, proving that the black color was the result of a superficial coating. It lacks the fire of ordinary diamonds, but gives brilliant metallic reflections, and exhibits, by transmitted light, the outlines of a black Greek cross. Something analogous to this, says Mr. Kunz, is seen at the Jardin des Plantes, Paris, and in the collection at Munich. The other diamond is a brilliant, apparently brown, but really giving out beautiful dark rose-red reflections, constituting a red and brown stone, or a red diamond with a brown cloud, the red predominating. The gem encloses numerous irregular-shaped cavities, either void, or filled with a transparent fluid.

What was called the "Cleveland gem," exhibited at the New Orleans Exposition, is a South African diamond, which weighed in its native state seventy-eight carats, and after cutting forty-two and one-half. It has a yellow tinge, is of great brilliancy, and exhibits a remarkable play of colors.

CHAPTER XIII.

THE PRECIOUS CORUNDUM. — SAPPHIRE, ASTERIA, EMERALD, AMETHYST, TOPAZ, RUBY.

It is an interesting fact that the rarest and the most valuable substances in nature are produced from the most common elements, — the diamond from carbon, and the gems of the corundum species from aluminum, one of the constituents of common clay. The corundum yields a larger variety of precious stones of the first rank than any other mineral. They are unaffected by chemical substances, their colors are the finest, and their hardness exceeds that of all others except the diamond, — qualities of great importance in gems.

A French chemist, less than a quarter of a century ago, prepared a metal on a commercial scale before unknown outside the laboratory, which, when combined with oxygen, forms alumina, and constitutes, in a pure crystalline state, the precious corundum known as sapphire, ruby, oriental emerald, oriental topaz, and oriental amethyst.

The first notice taken of corundum as of any scientific interest was in the last of the eighteenth century, by Sir Charles Greville; and though it was used by the nations of antiquity for dressing stones four thousand years ago, it has never been properly mined until recently, having been previously obtained in small quantities from surface-washings in Hindostan, Siberia, China, and some other places.

Col. C. W. Jenks discovered a remarkable deposit of this

mineral in Macon County, west of the Blue Ridge, North Carolina, in 1858, enclosed in the veins of a green rock, thought to be serpentine or lepidolite. Some attempts were then made, for the first time, at mining this stone, but it was not until 1872 that any systematic work was carried on at these mines. Since that date, a number of specimens of the precious corundum have been collected from this locality, which have been regarded by some as equal to those found in waterworn pebbles, and the compeers of the best oriental varieties; but skilful experts pronounce them inferior in quality to Asiatic specimens.

Two colors in the same crystal are rare even in the Ceylon corundums, but those of Macon County sometimes exhibit several hues in the same stone, — as red, pink, green, yellow, and others, and occasionally the colors blend, yet appear distinct when seen at different angles. Some of the crystals are of a large size, one having been found which weighed three hundred and twelve pounds, now belonging to the cabinet of Amherst College. Nine different varieties of the corundum designated oriental, have been taken from the mines since their opening. The associated gem-minerals are similar to those found with the Ceylon specimens, including chysolite, spinel, zircon, tourmaline, chalcedony, and rock-crystal.

Mr. Hamlin says, in reference to these corundums, that many of the crystals are limpid, but in consequence of their being crossed in all directions with cleavage-planes, and from the irregular distribution of their colors, they are unsuitable for jewelry, with the exception of very small gems of a few grains in weight, which may be cut from some transparent masses. He admits the colors are often very fine, especially the blues and yellows, but the reds lack the true "pigeon's-blood" tint. Mr. Kunz, writing on the same subject, says that of fifty specimens found at the Jenks mine, some of them weighing two carats, about one-half were of good color and possessed the characteristics of true gems, though none had a "higher value than, possibly, one hundred dollars." The other principal localities for sapphires in the United States, remarks this writer, are in the region near Helena, Montana, where they occur in sand collected in the sluice-boxes, during the process of mining for gold. The gems from this locality present quite a variety of colors, and are frequently dichroic, often blue in one direction and red in another, or blue and light green. Perfect gems are frequently met with weighing from four to nine carats. The value of the stones from this district reaches, at least, two thousand dollars per annum.

He mentions an interesting jewel belonging to Messrs. Tiffany and Company, made of these dichroic gems in the form of a crescent, which displays at one end red stones, and at the other blue, while the centre is composed of those affording different shades of bluish red. The entire crescent becomes red under artificial light.

The gem varieties of the corundum have always been regarded as the most valuable among precious stones by oriental nations, as they are at the present time by western races, unless the diamond be excepted. Of all the ornamental stones, the sapphire was the most highly esteemed by the ancients. It was the "gem of gems," the sacred stone par excellence, and the one most frequently consecrated to their divinities. The author of the Pentateuch, in describing the manifestation of Jehovah to the people of Israel, says: "There was under his feet, as it were, a paved work of sapphire stones, and, as it were, the body of heaven in his clearness." The sapphire mentioned in this quotation corresponds to the modern gem bearing that name, in its color and clearness (transparency), with something of the effulgence of the heavens. There is a legend among the Jews that the Ten Commandments were engraved upon this gem. According to Persian cosmogony, the globe rests on a vast sapphire, whose reflections give the blue to the sky.

Until the present century, it was thought the corundum and the sapphire were distinct minerals; and even at the present time writers differ in their methods of classification, some arranging under the generic term sapphire all hyalin corundums, making the ruby a red variety of sapphire, the emerald a green, and the topaz a yellow sapphire. Other mineralogists include the girasol, peridot, and aquamarine. The name is here applied to the oriental or true sapphire, — a variety of the precious corundum. What are called Brazilian sapphires, the French sapphires of Puy, and sapphire-d'eau, belong to other species. The Puy sapphires, of a fine blue color, are merely hyalin quartz.

The blue sapphire is supposed to be identical with the hyacinthus of Pliny; the white variety, with his adamas. Both its primitive and modern names refer to its color — azure. The origin of this epithet has been ascribed to a Syriac word, meaning "to shine," as well as to the Arabic "jacut," which may have suggested the term "huakinthos" to the Greeks. Though the ancients gave the name to all blue stones, yet their hyacinthus was always blue, never green, red, yellow or purple; and most modern mineralogists consider it identical with the sapphire of the present day. The description given by Solinus, a connoisseur in gems, who flourished two centuries after Pliny, corresponds to that applied to this stone. It came from the East, was of a cerulean hue, extremely liable to blemishes, and very hard and cold. In his day, it was obtained, as it is now, from Ceylon, where it was found in rolled pebbles mixed with gravel taken from the beds of streams.

It was used for jewelry in the Middle Ages, polished but not faceted, as may be seen in the crowns of Lombardy and Hungary, and the crowns of the Gothic sovereigns and nobles which have been recently discovered near Toledo.

Pure colorless sapphires are exceedingly rare, and sometimes are mistaken for diamonds, though the latter surpass them in fire and lustre. Sir David Brewster believed the white variety, on account of its structure and refractive power, was superior to all other transparent minerals for the lenses of microscopes. This gem has been known to disclose a different color by natural light from that seen by artificial light; that is, it may appear blue in the day, and purple in the evening, a phenomenon accounted for by supposing there exists an excess of red not visible by solar light.

The green variety of the precious corundum, termed oriental *emerald*, is one of the rarest gems in existence, and when of a lively green color, far excels the ordinary emerald in brilliancy and lustre. When it displays tints of sea-green or bluish green similar to those of the beryl, it receives the epithet of oriental *aquamarine.* Mr. Hamlin mentions some small, beautiful gems of this class obtained from the gold-fields of Montana.

The purple or violet corundum, called oriental *amethyst*, also very rare, seems to combine the hues of both the sapphire and the ruby. It is distinguished from quartz amethyst by its superior brilliancy, hardness, and beauty, though the latter is sometimes sold for the more valuable corundum variety. Some fine specimens are found at Dresden, and a few, with engravings, in the Vatican collection.

The yellow corundum, denominated oriental *topaz*, is more

plentiful than either the green or purple varieties, though it is seldom found without imperfections; but when free from defects, of fine color and perfectly transparent, it constitutes a very beautiful gem, rivalling the yellow diamond and the yellow zircon in brilliancy. Its commercial value is less than that of any other variety of the corundum species.

The remarkable coldness of the sapphire, due to its great density, gave rise to the notion that it would extinguish fire. It has been regarded one of the most appropriate gems for the episcopal ring of office, on account of its reputed character of preserving the virtue of the wearer, hence the oldest ecclesiastical jewel extant is set with a sapphire.

This precious stone possesses great refractive powers, a high specific gravity, and ranks next to the diamond in hardness; it crystallizes in six-sided pyramids and prisms. The color ranges from white to very deep blue, approaching a black tinted with red — a pure blue is said to be rare; but the most approved shade is styled "royal blue." Black sapphires are occasionally met with, but the hyacinthine tint is exceedingly rare, though one specimen, bearing a Greek engraving, is known to connoisseurs. The sapphire may be rendered perfectly colorless by heat, when it acquires great brilliancy, and closely resembles the diamond, being apparently inferior only in iridescence, a defect which the lapidaries of the Cinque-cento period remedied, to a certain extent, by artificial means.

Oriental sapphires are obtained from Ceylon, Pegu, Arabia, and some other parts of Asia; while others perhaps of equal intrinsic value, but less prized, are found in Siberia, Bohemia, Greece, Saxony, the Alps, France, Brazil, and the United States. Ceylon is by far the most productive region for this gem as well as for many others; the mines have been worked for many centuries, and have yielded innumerable specimens for the markets of the world. Its commercial value has been variable, the price being sometimes equal to that of the emerald, and at other times falling below it; stones of faultless tint command very high prices at the present time.

The sapphire is more abundant than the ruby, and in its native state occurs of a large size. Instances of this kind are seen in some of the musuems and collections of Europe, especially one specimen, of gigantic dimensions and great beauty, in Vienna, besides several others in the Green Vaults at Dresden, and in the imperial treasury of Russia.

Some blue stones of different species have occasionally been sold for genuine sapphires, but it is not difficult for an expert to detect the counterfeit. The iolite, which is one of them, may be known by its superior dichroism; kyanite, by its softness, while the blue tourmaline and the blue beryl, both rare varieties, may be recognized by other tests, but glass imitations are more deceptive to the eye, though readily yielding in hardness. The blue diamond is distinguished from it by superior hardness and brilliancy.

Among the celebrated sapphires is one belonging to the crown of Saxony, purchased, it is claimed, from an Afghan, the owner of the Orloff diamond, and considered the finest known, and two magnificent gems, owned by the Baroness Burdett-Coutts Bartlett, valued at nearly two hundred thousand dollars. The Lennox, or Darnley sapphire, now belonging to the Queen of Great Britain, was set as a heart-shaped pendant for Margaret Douglas, in 1575. This ornament, consisting of a gold heart more than two inches in diameter, and embellished with a sapphire, a ruby, and an emerald, is marked by three divisions—front, reverse, and interior, and combines numerous emblems and mottoes. A sapphire cut in the form of a rose, and once owned by Edward the Confessor, ornaments the centre of the cross in the royal crown of England, while a fine specimen, the purchase of George IV., is displayed upon the front. It was a sapphire ring taken from the finger of Elizabeth, just after she expired, which was sent to James VI. of Scotland as a token of his accession to the English throne.

It is reported that a remarkable specimen weighing nine hundred and fifty-one carats, of a beautiful blue tint and without a flaw, belongs to the royal treasury of Ava, Burmah; but as these jewels are guarded with jealous care, the statement must be accepted with caution, as the size is without a parallel.

An exquisite Indian sapphire, of one hundred and thirty-two and one-half carats, is seen in the collection at the Jardin des Plantes, Paris, while the British Museum is graced with a small statue of Buddha cut from a single gem of this kind. The Hope collection numbers among its treasures a specimen which exhibits varying colors under different kinds of light; and a sapphire owned by Louis XIV. presented the curious phenomenon of having a stripe of yellow topaz pass through its centre, which is said to have afforded Mme. Genlis the foundation of one of her novels. The most celebrated of the antique sapphires is the signet of the Emperor Constantius II., weighing fifty-three carats, and found in the Rinuccini collection.

Engravings on this gem before the Roman Empire are said to be exceedingly rare, but they occasionally occur, dating from the later empire, more especially during the fifth century. The pendants in the votive crowns of the Gothic kings, and the specimens in the Church of St. Ambrogio, Milan, are said to be of Indian origin. It is admitted that most of the engravings of the Renaissance, supposed to be on diamonds, are really on white sapphire or white topaz. A portrait of Matthew Paris engraved on sapphire is regarded by connoisseurs the most notable work of the kind during the Gothic period.

Some of the best known engravings on this stone are, probably, a head of Jupiter, in Greek style, set in the handle of a Turkish dagger; a head of Medusa, and another of Caracalla, in the Marlborough collection; the head of Julius Cæsar and of Apollo, formerly belonging to the Herz; a portrait of Pope Paul III., in the Pulsky, one of Henry IV. of France, and a cameo representing Hebe and the eagle, cut on a stone of this kind, measuring one inch and one-half by one and one-fourth.

The collection of the late Duke of Brunswick comprised a sapphire engraved with the arms of England, which formerly belonged to Mary Queen of Scots; a specimen bearing a female figure, and conspicuous for its dichroism, is found among the imperial jewels of Russia, and the French cabinet contains an intaglio in this gem with the portrait of the Emperor Pertinax.

Asteria. — This name was applied by the ancients to a variety of certain species of precious stones, more especially the corundum and the quartz, which, on account of a peculiar structure, displays divergent rays of light, like those of a star. This singular phenomenon is exhibited only in certain translucent or semi-opaque stones, cut *en cabochon*, or in hexagonal prisms, with the top rounded off. These rays are white or only slightly tinged, though the gem may be of various colors, and are most distinctly seen in sunlight or by the bright flame of a candle. The cause of this appearance is the numerous minute crystals arranged in different angles within the stone, which reflect the light so as to produce the stellar rays.

The corundums yielding this variety are called star-sapphires and star-rubies; when they assume a fibrous texture, they are called cat's-eye.

The ancients placed a high value upon the asteria, which they regarded as a powerful love-charm and for this purpose, according to tradition, it was worn by Helen of Homeric fame; so, then, to this beautiful gem were due all the calamities of the Trojan war. Very fine specimens of the star-sapphire are found in the collection of the École des Mines, and an extraordinary asteriated diamond at the Jardin des Plantes.

The term *girasol,* like asteria, is applied to certain gem-stones peculiar in structure, rather than as a variety of a particular species. The name, signifying "to turn to the sun," was given to it on account of the remarkably radiant light it emits when exposed to the solar rays, which moves as the stone is turned in different directions. The opal girasol possesses this quality in a higher degree than the sapphire specimens. The largest known girasol, called the Ruspoli, of one hundred and twenty-three carats weight, now in the Museum of Mineralogy, Paris, was found in Bengal and sold for thirty-four thousand dollars.

Ruby. — It is supposed the ruby corresponds to the *anthrax* of Theophrastus, and the *lychnis* of Pliny, one of the species of stones to which he gives the general appellation *carbunculus.* Both its Greek and Latin names were conferred in reference to some characteristic quality, as anthrax (red coal), in allusion to the color, and lychnis to its capacity of becoming very brilliant by lamplight. Its modern name ruby, rubino (red), is only an epithet for the red corundum, or red sapphire.

The ruby has the same chemical composition as the sapphire, pure alumina, with a difference of coloring matter, and ranks next to it in hardness. It possesses double refraction, exhibits electric properties, and, like the sapphire, its crystals are double six-sided prisms. The color varies from a rose to the deepest carmine, but the most approved tint is that of pigeon's blood; sometimes the same crystal exhibits different colors. There is only one true ruby, the oriental, of the corundum species, but the name has been applied to other gems, and we have the Brazilian ruby and the balas-ruby, which are not rubies at all, being different in composition and form of crystallization from the real ruby. The oriental carnelian, of a brownish red color, has been called a ruby, while some of the earlier mineralogists have classed under this name a dozen or more other gems of different composition and properties, which has caused no little confusion in the classification of precious stones.

A ruby of the finest color is one of the most beautiful and the most valuable of all the gems, and, compared to its size, the price is higher than that of any other, surpassing even the diamond, in the ratio of five to one, while there is no other that increases in commercial value so much in proportion to its increase in weight. It is found in Ceylon, the Burman Empire, British Burmah, Siam, Tartary, Bohemia, France, and on the Western Continent, but the best and most numerous specimens are obtained from a place sixty or seventy miles from Mandalay, the capital of Burmah, whose king is styled "Lord of the Rubies." These mines were a royal monopoly, and laws were in force strictly prohibiting all fine specimens from being carried out of the country, and to this cause is attributed the extreme rarity of large rubies in Europe. Strangers not being allowed to visit the region, very few Europeans have ever had access to the mines; consequently all knowledge of them is derived through government officials, who represent them as best suits the royal wishes.* When a valuable ruby was discovered, the occasion of taking

* These mines, since the annexation of Northern Burmah to Great Britain, will, probably, be accessible to any one who wishes to visit them.

possession of it was attended with great pomp and ceremony magnified by the imposing spectacle of a procession of grandees, accompanied by soldiers and elephants, despatched by the government to meet and escort the gem to court, like some distinguished prince or ambassador.

The Burman mines yield the sapphire, but the ruby is by far the most abundant; they are cut by native lapidaries, at Amarapura, not far from the capital, from which they are exported to Rangoon and Calcutta, the great gem-markets of the East. Mr. Streeter says, at the present time (1877), Pegu is the most prolific source of the ruby; there are also mines in active operation in the Chinese provinces bordering upon the Burman Empire, and, probably, in other places of China, but there is no doubt that Ceylon ranks next to Burmah for its rubies, and excels that country in the abundance of its sapphires. The Ceylon rubies are considered inferior in hue to the continental varieties, but surpass them in brilliancy, in consequence of being less opalescent; their violet tinge renders them less beautiful by daylight, but more desirable for evening ornaments, when they develop a fine prismatic red under the influence of artificial light.

The Cingalese, like other orientals, keep all the best specimens; and the lapidaries, after the fashion of their craft in other countries, manufacture these gems for the credulous purchaser.

The lychnis, the ruby of the ancients, was found in Asia Minor, but their best specimens came from India, where, at present, the production is very limited, and even in Tavernier's day the supply did not equal the demand. The famous ruby-mines at Badakshan, Tartary, were known to the Mogul emperors of India, from which these voluptuous rulers obtained at least a part of their abundant store. But the ruby is not exclusively an orientalist, since it has been discovered in the itacolumite of Brazil, and in various localities in the United States. It is a native of Australia, where the miners have given it the name of garnet, from misunderstanding its true character. There are other red gems which may easily pass for the ruby when judged by sight only, as the spinel and the garnet, detected by holding the stone up to the light, when it appears dark and opaque if it is a garnet, but if the gem is a true ruby it will be transparent and exhibit the conventional pigeon's-blood tint. Nearly all the great historic rubies now extant have been pronounced spinels by modern mineralogists, but is there not some doubt about the accuracy of this sweeping condemnation, except in instances where the nature of the gem has been subjected to the strictest tests? Rubies were imitated in paste by the ancients, with remarkable skill, as they are at present, even to their flaws. Parisian jewellers impart to a pale, valueless specimen the richest color by filling the inside setting with ruby enamel.

The rubies in ancient jewelry were polished but not often faceted or engraved, on account of the repugnance of the artists to the waste necessarily involved, and it has been supposed there were no antique engravings on this gem, but King says, though they are very rare, yet a few examples are known, and mentions as illustrations the head of Hercules, in the Webb cabinet; the head of Thetis with a helmet, a work of the Cinque-cento period, in the Herz collection; and a Bacchante crowned with ivy, in the Fould. An intaglio bearing the head of M. Aurelius, belonging to this writer, which had been considered a ruby, proved to be a balas. A ruby engraved with the names of several Indian kings was owned by Runjeet Singh, and one in the Persian treasury, described by Chardin, a traveller and dealer in gems, is represented to have been of the size of a half hen's egg and bearing an inscription. A pink ruby used as a signet by one of the Persian kings was engraved with the motto, "Riches are the source of prosperity," while another seal of the same kind of gem bore the inscription, "Splendor and Prosperity."

The best recent engravings on this stone are a head of Louis XII.; a fine specimen, belonging to the Queen of Great Britain; the head of Henry IV. of France, with the date 1598, in the Orléans collection; a Venus Victrix, an Osiris, and a Gorgon's head, among the Devonshire jewels. The most approved form of cutting is the half-brilliant.

The number of fine large rubies of undoubted genuineness is small, few even of Indian origin exceeding twenty-four carats, though there are on record several gems of immense size reputed to be rubies. The largest of these are owned by Asiatic princes, while those belonging to the crown jewels of European sovereigns, those of Russia, perhaps, excepted, are generally small or of ordinary size. The large gems in the English crown regarded as rubies have been suspected of being spinels, and the same doubt has been raised in reference to those of some other countries. The fine ruby cut in the form of a dragon, of only eight and three-sixteenths carats, which adorns the Golden Fleece belonging to the French crown jewels, is one of the largest in Europe, according to King, *decidedly* known to be genuine, though this collection is said to comprise a specimen of seventy-three and three-sixteenths carats. Probably the finest collection of rubies, including some of vast size, belongs to the imperial treasury of Russia. The specimen presented to the Empress Catharine II., claimed to be as large as a pigeon's egg, if it is genuine, is the largest ruby, without doubt, on the continent. Several of fine quality and considerable size are found among the crown jewels of Austria.

Tavernier mentions a ruby, owned by the Shah of Persia, which equalled a hen's egg in magnitude, and was bored through the centre. The large specimens which he describes as ornamenting the thrones of the Indian princes, as well as the immense ruby of General Wallenstein, obtained from the Bohemian mines, were balas, it is believed, and not true rubies.

Several others of immense size and marvellous beauty,—whether genuine or counterfeit has not been fully established,—are known in the literature of gems. The Devonshire ruby, though small, weighing only three or four carats, is considered the paragon among these precious stones for the beauty of its color; this exquisite little gem is engraved with the figure of Venus and Cupid.

The royal treasury of the Burman Empire is believed to hold a remarkably large and beautiful ruby of immense value, but as it has never been seen by any European its character has been suspected. A ruby of great size cut as a Chinese idol captured at the sacking of Peking by the French, and purchased by the Duke of Brunswick, was valued at three thousand dollars, a very small sum for a genuine specimen. The jewels of Charles the Bold, lost at the battle of Grandson, comprised three rubies called the "Three Brothers," which have been thought to be spurious, besides two others, named "La Hotte" and "La Belle de Flandres." James I. of England mentions a jewel known as the "Three Brothers" which may have been the same as the one lost by the Duke of Burgundy. The Herz collection included a necklace of rubies and emeralds linked together by twisted gold wire; while Rudolph II. of Austria owned a ruby of gigantic size which had been purchased for one hundred and fifty thousand ducats, and bequeathed to him by his sister, the queen-dowager of France.

The two most important rubies ever brought to Europe,

it is said by Mr. Streeter, were purchased from the Burman government in 1875. Their exportation from the country caused considerable excitement among the populace, as there is a decided opposition to having any fine stones leave the kingdom. One of these gems, displaying a fine, dark color, weighed thirty-seven carats; the other, a drop-shaped stone, weighed forty-seven and one-sixteenth carats; but by cutting they were reduced to thirty-two and five-sixteenths and thirty-nine and seven-sixteenths carats respectively. No European regalia, it is added, contain two gems of this kind of so fine a quality, yet they were allowed to pass into the hands of parties abroad.

The ruby has always been highly esteemed in oriental countries, where it was employed as a metaphor by the sacred writers for what was most valuable and excellent in moral attainments; it holds also a conspicuous place in ancient romances as well as eastern legends. The Greeks invented many curious stories about the carbunculus and the lychnis; one of these legends relates that a lame stork, as an expression of gratitude for the kindness it had received from Heraclea, placed in her lap a wonderful gem, supposed to be the ruby, which lighted up her room at night by its marvellous brilliancy. The lychnis or lamp-stone worn by the goddess Astarte made her temple luminous at night, by its supernatural lustre.

There were writers in the sixteenth century who saw as many marvels as the old heathen romancers; for instance, those who witnessed the brilliancy of the chrysolampis dedicated to Lady Hildegarde, wife of Theodoric, Count of Holland, which lighted the chapel by night.

So famous a traveller as Sir John Mandeville of the fourteenth century says: "The Emperor of Cathaye (China) has in his chamber a ruby and a carbuncle half a foot in length supported by pillars of gold, and of such brilliancy that they make the night as luminous as the day." But Epiphanius, who lived in the fourth century, bears off the palm, since his carbuncle could not be concealed by any covering whatever, its brilliancy was so penetrating. Catherine of Aragon, says a contemporary, wore a ring set with a stone, thought to be a ruby, luminous by night, while as late as the eighteenth century this gem was supposed to give a warning of misfortune to the owner, by a loss of brilliancy and change of color.

The traditions about the luminous property of the ruby and some other gems may be traced to a well known quality of the diamond, — phosphorescence,—the only precious stone in which it inheres; but in an age when the illusions of the imagination had not been dispelled by scientific experiment, it was easy to ascribe this quality to other gems. This peculiar phenomenon of the diamond, it has been observed, was undoubtedly noticed when persons wearing large numbers of them passed from the blaze of a tropical sun, to the comparative darkness of oriental rooms, a circumstance which afforded some foundation for the marvellous tales related about the properties of gems.

CHAPTER XIV.

THE BERYL.

The name of this mineral, from the Persian *belur*, Latin *beryllus*, is applied to a species including several varieties,— the emerald; aquamarine; Davidsonite, a greenish-yellow beryl, found near Aberdeen, Scotland; and Goshenite, so called for Goshen, Massachusetts, a place affording crystals of gigantic size. The beryl, in some of its varieties, has been discovered in various localities in both hemispheres. In this country, it occurs in different parts of New England, in Pennsylvania, North Carolina, and other States. Fine crystals have been found in Royalston, Massachusetts, displaying a great variety of colors, comprising different shades of green, light and deep yellow, the gold tint of the topaz, sherry-wine, and a clear blue, approaching that of the sapphire in its purity; while those from Fitchburg, in the same State, resemble the topaz and the chrysoberyl in color and hardness. Crystals of large size, occasionally measuring several feet in length, are known near Stoneham, Maine, but not usually of a quality to yield gems, though a few remarkable examples have been obtained from this locality, considered equal to the best foreign beryls. The largest specimen, of a rich sea-green or a sea-blue, according to the direction in which it was viewed, would, when cut, afford a gem of thirty carats weight. Fine specimens have been developed in Colorado.

Beautiful blue beryls may be obtained from the Mourne Mountains, Ireland, but the largest number of superior quality are, undoubtedly, brought from the Russian Empire, principally from Siberia and the Urals, where the ancient Romans probably found them. Specimens from these regions display a great variety of splendid hues, including green, blue, white, yellow, and pink, — a rare color in this gem.

Beryl crystals not unfrequently attain a colossal magnitude. A specimen owned by Don Pedro, of Brazil, shaped like the head of a calf, weighed two hundred and twenty-five ounces Troy, but this crystal is almost microscopic compared to some obtained from Grafton, New Hampshire, seen in the collections of different museums. One single example taken from this locality yielded the astonishing weight of two thousand and nine hundred pounds, avoirdupois, while a second crystal gave a weight of one thousand and seventy-six pounds.

The beryl occasionally exhibits two distinct colors in the same specimen, but generally they are monochromatic, passing into white at the extremities. The tints are considerably varied, embracing shades of green, blue, yellow, and rose, due chiefly to iron, except in the emerald and aquamarine, which are supposed to be the result of chrome. It always occurs in crystals, from transparent to opaque, which assume the form of six-sided prisms, and are sometimes striated. The yellow beryl has been called chrysolite, but it differs from that precious stone in everything except color. The white or colorless variety often passes by the name of "Rhine diamond" — which is really glass — on account of its fine lustre, and it is sometimes taken for the white topaz.

The beryl and the emerald were formerly classed as distinct species, but most modern mineralogists call the fine, transparent green beryl emerald, and the paler tints aquamarine.

The ancients obtained the beryl at first from India, then from Arabia, and later from Siberia. The Indian lapidaries were accustomed to heighten its brilliancy by perforations, and also substitute imitations in colored rock-crystal for the genuine article—a practice which suggests the scarcity of oriental beryls. The Greeks used this precious stone for engraving more than two thousand years ago, and the Romans followed their example at a later period, though they used the emerald more frequently for this purpose. King says the beryl was the only gem cut in facets by Roman artists.

There are few ancient intagli on beryl, though they frequently occur of the Renaissance and later periods. The finest engraved specimen is the Glaucus, or Palæmon on a dolphin, in the Mertens-Schaffhausen collection, while one of the chief ornaments of the Cracherode gems, in the British Museum, consists of a beryl engraved with a Cupid on a dolphin.

One of the finest known beryls, belonging to the Hope collection, is an Indian gem, weighing six and one-half ounces, and is valued at two thousand three hundred and twenty-five dollars. A magnificent blue beryl surmounts the globe in the royal crown of Great Britain.

Mr. Hamlin refers to several remarkable gems of this species, including a specimen from the Siberian mines, of yellowish green and greenish blue tints, constituting one of the most beautiful known; and another from Ireland, belonging to the Vaux collection, in Philadelphia, which measures nine inches in length and six in circumference, of a rich green, with the peculiarity of being transparent at one end of the crystal, while the remainder is only translucent. A Siberian beryl in the possession of Mr. Clay, of Philadelphia, measuring two inches in diameter, displays a cerulean blue externally, but is of a decided green in the interior.

Aquamarine.—This epithet, applied to a variety of other

mineral species besides the beryl, is given to precious stones which resemble in color the water of the sea. The beryl aquamarine is light blue or sea-green, and, though of less value in commerce than the emerald, it possesses the quality of retaining its brilliancy by candle-light, a merit which does not belóng to many of the more costly gems, consequently it is a favorite in the fashionable circles of some countries. It is held in England, says Streeter, in the same high estimation that the topaz is in Spain.

Aquamarine occurs in many different localities, but the greater part used for jewelry is obtained from India, the Ural and the Altai Mountains, and Brazil. It has sometimes passed for other gems, and is so near the color of greenish glass that frauds have been easily perpetrated by throwing fragments of green bottles into the sea to be washed ashore and gathered as pebbles of aquamarine.

A superb specimen discovered in Russia, in 1827, was valued at the marvellous price of one hundred and eleven thousand six hundred dollars. The historical aquamarine which once adorned the tiara of Pope Julius II., having passed into the control of the French, was placed in the Museum of Natural History in Paris, where it remained for more than three centuries, until it was returned to the Vatican by Napoleon I., who presented it to Pius VII. It is described as a beautiful sea-green gem more than two inches in length, and between two and three in depth.

The hilt of Murat's sword, now in the South Kensington Museum, was ornamented with one of these stones, weighing three and one-half ounces. It is said that the Emperor of Brazil owns a large, splendid aquamarine, without a flaw and of remarkable transparency. The finest gem of this variety of beryl ever discovered in the United States, says Mr. Kunz, was obtained from Stoneham, Maine, which, cut as a brilliant, weighs one hundred thirty-three and three-fourths carats. With the exception of a few hair-like striations in the interior of the stone, it is of a perfect bluish green color.

This variety of the beryl has been employed both in ancient and modern times for engraving. A famous intaglio cut from this gem, presented to the Abbey of Saint Denis by Charlemagne, and employed to adorn a gold reliquary, now deposited in the National Library of Paris, is engraved with the portrait of Julia, the daughter of the Emperor Titus.

Emerald. — Probably no other precious stone has been the subject of so much exaggeration, especially among early writers, as the emerald, and none has been more highly prized, both in ancient and modern times, for its beauty and excellence as a gem-stone. Its pure tint is unsurpassed by that of any other object in nature, and when transparent and entirely free from blemishes, it constitutes one of the most desirable ornamental stones known to the lapidary.

This gem has generally been selected to represent marine subjects, both in art and literature, a use which has, perhaps, some connection with its ancient names "*smaragdos,*" Greek, "*smaragdus,*" Latin, and "*marakata,*" Sanskrit, — all having some reference to the sea.

The composition is nearly identical with that of the beryl, — silica, alumina, and glucina, — but different from the corundum variety which is pure alumina; its lustre has been compared to the sheen of olive-oil. If held so as to reflect the light, it appears to be silvered on the back, and its green color will disappear when the plane is brought to a particular angle, a peculiarity not observable in any other gem. When struck by a hammer, the crystals will break across the prism, yielding slices with smooth and brilliant faces; many of the oriental emeralds consist of these layers mounted without any artificial cutting or polishing. When seen at right angles to the optic axes, the emerald presents one image of greenish yellow and another of greenish blue tints.

The color of this gem, which constitutes one of its principal charms, is a lively grass-green without admixture of any other hues, and has generally been supposed to be due to the oxide of chrome, though this theory is not universally adopted; some scientists referring it to copper, some to iron, and others to an entirely different cause. It is extremely difficult, says Mr. Rudler, to determine the precise nature of the coloring matter present in gems, since it is so intense that the smallest possible quantity is sufficient to give them a decided tint.

This question gave rise to some very interesting experiments made by M. Lewy, in 1848, which led him to conclude the color was derived from some organic matter similar to the coloring substance of the green leaves of plants, called chlorophyl. He assumed that the emerald loses its hue by heat, while chromic oxide is a stable pigment, and ought not to be affected by it. This experimenter burned it in oxygen, and found that carbonic acid was produced, as when a diamond is subjected to the same operation; therefore he believed it must contain carbon. The conclusion reached after repeated experiments was that the coloring substance consists of a compound of carbon and hydrogen, resulting from organic matter. Emeralds from Muzo, South America, have been found in a fossiliferous limestone of a bituminous nature; therefore their color, it is reasoned, must have been caused by the decomposition of animal matter, similar to that found in vegetables. M. Lewy obtained by his analyses a certain per cent of water, which led him to infer that these gems were formed by a chemical solution. These interesting theories have, of course, their objectors, who argue that the Muzo emeralds do not lose color except by exposure for several hours to red heat and a prolonged state of fusion, which no organic matter could possibly endure without destruction. Finally, since M. Lewy's reasoning is supposed to be illogical, the chromic origin of the coloring agent has been taken up again, if it ever was abandoned, by a certain school of scientists.

The localities where the emerald occurs are various and widely separated. Those of the finest color are found at the Muzo mines, near Bogota, New Grenada, in a calcareous rock, in isolated crystals and geodes. They have also been obtained from other regions in South America, from Siberia, the Deccan in India, Egypt, the Tyrol, France, Norway, and North Carolina, in the United States. It is probable the emerald was a native of Mexico, since it was used very abundantly for ornaments by the aborigines of the Western Continent, at the time of the Spanish conquest. Peru was the great storehouse where it was obtained for more than two centuries after the discovery of the land of the Incas, and immense quantities were imported into Europe, as we learn from D'Acosta, one of the writers of the times, who mentions a ship coming from the western world which had on board two chests of these gems, the spoils of war, each weighing one hundred pounds. The soldiers of Pizarro, not understanding the properties of the emerald, destroyed many fine specimens in their clumsy experiments to test its frangibility.

The natives, who had been acquainted with the use of this precious stone from time immemorial, venerated it as the abode of their favorite divinity. The chief goddess of Peru, says De la Vega, was an emerald as large as an ostrich egg, and the principal offerings made to it consisted of this gem, of which the native priests were the fortunate guardians. Many of these costly gifts fell into the hands of the conquerors, but the great emerald goddess was spirited away and her hiding-place could never be discovered.

A slight variation from the above tradition makes the emerald only the dwelling-place of the goddess Esmeralda, and not the veritable divinity herself. The Peruvians, like the nations of the Eastern hemisphere, cherished the belief that mines of the precious metals and precious stones were guarded by demons and griffons.

Emeralds of great value, and in large quantities, were carried off by the Spanish brigands, during their invasion of Mexico, many of which found their way into the royal treasury of Spain, while others were retained by the conquerors. Five, of remarkable beauty and of curious design, given by Cortez to his bride, were the cause of his loss of the royal favor, as stated by his biographer; consequently, the origin of the misfortunes that befell him in his last years. These emeralds are described as marvels of the lapidary's skill, one having been cut in the form of a rose, another in that of a horn, a third representing a fish with golden eyes, the fourth a bell with a tongue of pearl, and the fifth a cup resting on a gold foot with small gold chains attached. Two of these gems bore inscriptions, and the whole set was valued at several million dollars. Cortez was offered a large sum for them, but he had the imprudence to refuse to dispose of his treasures, even to Charles V., who wanted them for the empress. This disregard of the imperial will was followed by a withdrawal of court favors, which, in those days, was a serious calamity. These coveted jewels were subsequently lost at sea during the shipwreck of the owner, on the Barbary coast, in 1529; still, he had other valuable emeralds left, which proved the wealth and luxury of the Montezumas, including two vases cut from this gem, priced at three hundred thousand ducats. Cortez, while in Mexico, sent to the emperor, as a present, an emerald pyramid with "a base of the size of a man's palm," besides other gifts, which were captured by the French, and went to enrich the collection of his rival, Francis I.

Charles received from Montezuma and the Spanish commissioners magnificent gifts of emeralds, pearls, and red gems supposed to be rubies, with two necklaces comprising from three to four hundred emeralds, — a bonus sufficient to satisfy a less ambitious prince than the German emperor.

After the conquest of the New World, emeralds became very plentiful in Europe, where before they were comparatively scarce. As a proof that the American variety had been adopted for the favorite ornament in the highest social circles, Hamlin refers to a parure made of remarkably beautiful specimens of this gem, which was bequeathed to her daughter by the Queen of Navarre in 1572. The Dresden Museum contains a large uncut emerald, the gift of Rudolph II., and the collection at Munich several of large size, from Peru. An emerald taken from the tomb of Charlemagne, which had been used by this conqueror as a talisman, came, by some unexplained fortune, into the possession of Aix-la-Chapelle, and was presented by the citizens to Napoleon I., who gave it to Queen Hortense, after having worn it at Austerlitz and Wagram.

The treasury of the Czar of Russia contains many fine emeralds, including some of large size and others of extraordinary beauty; to the latter class belongs a gem of thirty carats, perfectly transparent, immaculate in color and considered one of the most superb in Europe. The crown of Vladimir, the state sceptre, the imperial orb, and the sceptre of Poland, preserved in the Kremlin, are more or less ornamented with emeralds, some of which were, undoubtedly, taken from the Siberian mines and may possibly be green tourmaline or the splendid green garnet found in this region.

Emeralds of surpassing beauty are said to be found in the rich collection at Constantinople; one faultless gem, weighing three hundred carats, is set in the handle of a poniard; another, whose genuineness has been questioned, weighs one hundred and twenty-five ounces, Troy.

The uncut Devonshire emerald, taken from the mines of Muzo, measures two inches in length, and weighs eight ounces eighteen pennyweights, but unfortunately its intrinsic value is greatly marred by flaws. A cluster of these gems, each more than an inch in diameter, perfect in color and brilliancy, and imbedded in white limestone, constituted one of the votive offerings to the celebrated shrine of Loretto, presented by a Spanish ambassador to Rome. Some of the Indian princes are the owners of valuable emeralds which they display upon their persons with other gems on certain occasions, as was related in the public journals when the Prince of Wales made the tour of their country. An emerald of a very large size was presented to the Queen of Great Britain by the Sultan of Oude, while a specimen of this kind, owned by Duleep Singh, is larger than the Devonshire. An emerald of the size of a walnut, engraved with the names of the kings who had owned it, is comprised in the Persian royal treasury.

The collection of precious stones at Madrid affords many emeralds of distinguished size, comparatively exempt from flaws, an occurrence so unusual that the expression "an emerald without a flaw" has passed into a proverb, to denote unattainable perfection. The French, during their invasion of the Spanish peninsula, carried off, with other precious stones, many of these fine emeralds, notably the one which was the most glorious gem in the crown of the Virgin in the Cathedral of Toledo.

The Prussian, Saxon, and papal crowns contain emeralds of remarkable size and beauty; Austria possesses one of immense proportions, said to weigh two thousand carats. There is a fine specimen belonging to the Townshend collection, which measures nearly half an inch across, and another, owned in London, possesses a magnitude of two and seven-eighths inches by two and five-eighths.

The famous emerald-mines of Mount Zebarah — "Mountain of Emeralds," — between the Nile and the Red Sea, mentioned by Strabo and other ancient writers, have recently been discovered by M. Caillaud, who found them nearly in the same condition in which they were left by the engineers of the time of the Ptolemies. They had been excavated to a great depth, and contained lateral passages, causeways, and other appliances of modern mining, together with various implements employed in the works lying about. Strabo speaks of them as mines of emeralds and other precious stones extracted by the Arabs.

M. Caillaud, deputed by the Viceroy of Egypt to reopen the mines, discovered that they had been worked to the depth of eight hundred feet, and that a part of them afforded space for four hundred workmen at once. Very extensive quarries were found seven leagues from Mount Zebarah, comprising more than one thousand excavations. The emeralds were deposited in a black, micaceous clay-slate penetrating the mass of granite, and sometimes in granite and hyalin quartz: they were of a pale green color and full of flaws. They are used in Cairo and Constantinople for jewelry and for decorating the imperial equipages.

The emerald has been discovered in the Ural and the Altai Mountains within the present century. The first stone found in the Urals was in 1830; a second was discovered the following year weighing ten and one-fourth carats; since then they have been mined to a considerable extent. It is very probable that the ancient Scythians obtained their supply from these mountains.

Peru, after the Spanish conquest, supplied these gems for the European markets until the mines were abandoned for the more recent fields of New Grenada. This state leased the gem-producing territories for a certain sum per annum, until there were no bidders for the privilege, not from any exhaustion of supply, but in consequence of the diversion of labor and capital to the gold regions; consequently, work at these emerald-mines was suspended for a long period. The French resumed the mining operations about the middle of the present century, and since then the best emeralds have been exported to Paris from New Grenada, where, during the Empire, they became extremely fashionable, green being the imperial color.

Emeralds were discovered in North Carolina in 1880, simultaneously with hiddenite, a variety of spodumene, by Mr. W. E. Hidden, associated with several other minerals, many of them constituting gem-stones. The emeralds were found, after persistent mining, at the depth of more than fifty feet, in veins or pockets of a rock resembling gneiss; sometimes these deposits were very close together, and contained only emeralds and hiddenite, and sometimes they comprised a variety of other minerals. The largest emerald crystal measured eight and one-half inches in length, and weighed nearly nine ounces. The greatest number found in one pocket was seventy-two, some of the specimens having a length from two to five inches, but the larger part were very small though of the finest tint, resembling the pure deep-colored gems of New Grenada. As an ornamental stone, the North Carolina emeralds have little value, says Mr. Hidden, but for cabinet specimens their prices range from twenty-five dollars to one thousand each.

The quality and tone of color vary in these gems, according to the different localities from which they are obtained. Those from New Grenada are considered the finest; the Siberian rank next, if they do not equal the South American specimens; the Indian and African stones are pale in color and full of flaws, while those found in Europe hold the lowest rank.

The phrase "Emerald Isle," applied to Ireland, is said to have come into vogue, not on account of its remarkable verdure, as is generally supposed, but from a circumstance connected with the gem itself. Pope Adrian sent to Henry II., of England, a ring set with an emerald, as the instrument of his investiture with the dominion of that island, which may aptly be compared to this gem set in the sea. This ring, which has disappeared from the royal archives, if it could be found, would be an object of curiosity and interest as the record of an historical event of important political results.

Comparatively few emeralds are engraved, partly on account of their brittleness and partly from their intrinsic value; therefore, antique intagli in this gem are exceedingly rare, and very few of these are earlier than the time of the Emperor Hadrian. The best examples known to antiquaries are an emerald engraved with the head of this emperor, another with the portrait of Sabina, his consort, and a third with the heads of both on the same stone. His patronage of the Egyptian system of mythology is shown by a remarkable intaglio with the head of the Solar Lion, a work of this period; in fact, a large portion of antique engraved emeralds were executed during the reign of this emperor, though the Etruscans, at a very early date, engraved this gem with scarabei.

Three engraved emeralds of antique workmanship belong to the Fould collection in Paris. The Devonshire parure has one, with a Gorgon's head in high relief; and an amulet formerly comprised among the Praun gems bears the head of Jupiter, with a serpent and crocodile, surrounded by the emblems of the planets. The last-named collection includes a Gnostic legend, containing several lines, cut on emerald. Engraved gems of this kind of precious stone were supposed to be endowed with remarkable mystic powers, and on that account were employed for amulets. Those with the representation of the eagle or the beetle were thought to be powerful agents in conciliating royal favor. A jewel engraved with the name of the Emperor Jehangir, and used as a signet, consisted of two emerald drops and two collets of rose diamonds with ruby borders, mounted in oriental fashion. It was presented to the East India Company by Shah Soojah, and purchased by Lord Auckland when Governor-General of India, and is owned at present, it is said, by the Hon. Miss Eden. Another remarkable engraving, representing the heads of the Apostles Peter and Paul, and Pope Benedict, was executed on a large table emerald, by one of the Costanzi.

There is unequivocal evidence that the ancients were well acquainted with the emerald, notwithstanding the attempts which have been made to prove they were not known to the Eastern world before the discovery of the Western continent. No emerald-mines, it has been said, are known in India, yet it was used by the natives for ornament; and, reasoning from these premises, Tavernier concluded they must have been introduced from America. There is, however, no doubt that the emerald was known both in Asia and Europe long before the Spanish conquests. They are mentioned by ancient writers and are known to have existed in collections, and to have been employed in antique jewelry, long before the fifteenth century. Examples are afforded in the tiara of Pope Julius II.; in the Iron Crown of Lombardy, presented to the Cathedral of Monza, 589 A. D.; in the Gothic crowns found near Toledo; in the Crown of Hungary, made in 1072, and in the crown of Edward the Confessor; they have been found in Etruscan tombs and at Herculaneum, made into necklaces. The Egyptian mines of the true emerald had been worked ages before the time of Theophrastus, who was contemporary with Plato, and it is quite likely those of Upper Egypt became an article of commerce to the people of India and to the earlier Greeks and Romans.

The mines of the Thebaid and of Ethiopia, in successful operation as late as the fourth century, have been exhausted or neglected. Emeralds have been mentioned by Heliodorus of the second century; Claudian, of the fourth, sings of

> "Breastplates of shining green with emeralds bright,
> And helmets rich with precious sapphires dight."

Ben Mansur, writing in the thirteenth century, refers to the emeralds on the borders of the land of the negroes.

There is always more or less perplexity about identifying the precious stones employed by the ancients, arising from their method of classification, as in the case of the emerald, a name applied to other gems, perhaps to all of a green color. But it is believed by the most experienced antiquaries that they were familiar with the genuine emerald; it was undoubtedly included in their smaragdus. Pliny enumerates twelve different kinds, two of which were probably identical with the modern emerald.

Theophrastus refers to a variety of this precious stone used for ornamenting gold vessels, the color harmonizing well with that metal, a quality belonging to our emerald. The Persians used it to embellish their goblets, a practice adopted by the Romans.

The emerald mentioned by Ezekiel as an article of merchandise in the Tyrian fairs may have been that stone, or turquoise, which is still mined at the foot of Mt. Sinai, or some other green gem. The jasper of the New Jerusalem, a "most precious stone," combining the green of the jasper with the transparency of the crystal, may have been the emerald.

In Pliny's account of precious stones, it is stated that the smaragdus was found in very large masses. Theophrastus says it is scarce and of small size, unless we credit the commentaries of the Egyptian rulers, which relate that a specimen sent to the King of Babylon was four cubits in length and three in breadth, and that the obelisk in the Temple of Jupiter, forty cubits high, was made of four emeralds. It is believed, at the present time, that these and other gigantic specimens were glass, as Alexandria, in Egypt, was noted for its glass manufactures. It is possible that some of these enormous emeralds were beryl, since crystals of the size of those from Grafton, New Hampshire, could have answered the purpose. Theophrastus speaks of a "bastard emerald" found in the copper-mines of Cyprus, which very likely was malachite or chrysocolla, or, as Hill suggests, rock-crystal tinged with green. He says the true emerald seems to be a production of jasper, giving as proof that a certain crystal known to him was half emerald and half jasper. The prase, or plasma, has been called the "mother of emeralds"; and it is possible the jasper, in some of its varieties, may have constituted the matrix of this gem.

The emerald was used in Pliny's time, as it has been since, to rest the eyes after overstraining them on difficult work, and it was also employed as lenses for near-sighted persons, and for mirrors. The Emperor Nero, who was very near-sighted, endeavored to remedy this defect, when witnessing the gladiatorial combats at Rome, by using a lens cut from this precious stone. Historians mention this fact in their descriptions of this notorious despot, corroborated by his portraits, in which the eyes are remarkably full, indicating myopy. If it was used for that object, it was quite probable it was hollowed out to serve as a concave lens and not as a mirror to reflect distant views. It is said, however, that he used this gem to reflect the images of any lurking assassin, a danger to which he was imminent.

To prove that the emerald has served the purposes of a reflecting body a story is told of the Emperor Maximilian II., who, on his visit to Ratisbon, when presented with a gold cup full of ducats, detected one of his courtiers helping himself to the contents, by the reflection of the scene in the emerald of the ring upon his finger.

The Hindoos have always valued the emerald very highly for ear-pendants and bracelets, which they drill and string as beads. These perforated gems are cut in two when used by European lapidaries. Tavernier testifies that all East Indians who could afford it wore in their ears a ruby and an emerald strung between pearls.

False emeralds were manufactured in the time of Pliny from rock-crystal and other inferior stones, by plunging the heated mineral into verdigris dissolved in turpentine. They were also imitated very successfully in glass, an art practised at the present day. The emerald column in the Temple of Melkart, at Tyre, which excited the wonder of Herodotus, was probably a shaft of glass made to enclose a lamp, thus leading the credulous to believe it shone by its own inherent brilliancy.

The substances most closely resembling the emerald, are green jasper, green spinel (laal), and green glass. Some remarkable instances of fraud have been practised by passing off articles made of glass for emerald, as the traditional Sacro Catino, "Sacred Cup," belonging to the Cathedral of Genoa, supposed to be emerald for centuries, is now believed to be glass. This cup, fourteen inches wide and five deep, is claimed to be the identical one used by our Saviour at the institution of the Lord's Supper, and once belonged to the banqueting plate of King Herod. It was given to the Republic of Genoa in 1101, as an equivalent for money due from the Crusaders who had captured it during their wars in the East. It was pawned for a large sum, nearly two hundred thousand dollars, in the beginning of the fourteenth century, but was redeemed under the belief that it was genuine emerald. When the French captured Genoa, they tested this famous relic and found it to be glass. A rival to the Genoese vessel, though of much smaller dimensions, was discovered at a monastery near Lyons in 1565.

A gigantic emerald, weighing twenty-nine pounds, given by Charlemagne to the Abbey of Richenau, could not stand the test of modern experimenters, who very unceremoniously pronounced it green glass. The most disappointing results followed a similar trial of the celebrated "Table of Solomon," discovered by the Arabs in the Gothic treasury of Toledo. It is described as a single piece of solid emerald encircled by three rows of pearls and supported by three hundred and sixty-five feet made of gems and massive gold. This table was plundered from the Temple of Jerusalem by Vespasian, and deposited in the Temple of Concord at Rome, but when the latter city was sacked by Alaric this relic formed a part of the spoils, and, like the Sacro Catino, it turned out to be glass.

Antique glass emeralds, far superior in color, lustre, and hardness to modern pastes, have sometimes been recut and faceted for rings, by modern Roman dealers, and sold to incautious dilettanti for the real article. The Cingalese are charged with using the bottoms of wine-bottles from which they cut "bona fide" emeralds for foreigners, and the "Brighton emeralds" purchased by visitors have a similar origin.

It is related that the tomb of one of the princes of Cyprus placed near the sea was surmounted by a lion with emerald eyes so lustrous that the fishes were scared away by their brilliancy, and, for the sake of the fisherman's craft, they were removed. It is possible, remarks Mr. King, that the marble lion brought from Cnidos with eyes deprived of their pupils, now in the British Museum, may be the identical lion of Cyprus.

Topaz.—The topaz of antiquity, it is supposed, was the chrysolite, while the modern topaz was known to the ancients by some other name ; that it was employed by them as a gem-stone is evident from Greek intagli, of a very early period, which have come down to the present day. The origin of the word has been ascribed to topaza, signifying "to guess," and also to Topazos, an island in the Red Sea, from which the early nations obtained their topaz, whatever that may have been. It is thought the greenish yellow topaz was called by the ancient lapidaries chrysoprase, a term now applied to a variety of quartz.

There are two kinds known to modern mineralogists ; the oriental, constituting a variety of the precious corundum, and sometimes denominated yellow sapphire; and the occidental, of a different composition, forming a distinct species.

This precious stone is one of the very few containing fluorine, Dieulafait says the only one, but some other writers give this element as one of the constituents of tourmaline.

The crystals assume the geometrical form of prisms with only one end terminating regularly, and exhibit distinct pleochroism, double refraction, and strong electric powers. The topaz holds the rank of eight in the scale of hardness, and exhibits greatly diversified colors, including different shades of yellow, gray, blue, rose, pink, red, green, citron, and a white, or colorless variety; the last often passes for diamond on account of its great brilliancy, as in the enormous Braganza or Portuguese gem generally supposed to be a topaz. The "Minas Novas," a white Brazilian topaz, so called from the province where it is found, is sometimes sold for diamond.

This species of precious stone is found in India, Siberia, Australia, Saxony, Austria, Brazil, Mexico, and in Maine, Connecticut, North Carolina, New Mexico, Colorado, and Arizona, in the United States. The Colorado specimens — one having been discovered weighing more than thirty carats,— are of a beautiful light blue color; the yield at Pike's Peak, as stated in Leslie's Magazine, exceeded at the time of writing one hundred dollars per annum. Though found in large crystals, only a small portion of these specimens are suitable for jewelry. Beautiful varieties occur in Siberia, saffron-yellow in India, wine-colored and pale violet in Saxony, a sea-green sometimes called aquamarine, in Bohemia, and blue in Scotland, while Brazil furnishes specimens in gold, ruby, rose, sapphire-blue, and light blue colors. The fine, delicate, sherry-colored stones from Siberia soon fade in the light, and on this account such specimens in the British Museum are kept covered.

The gem known as Brazilian sapphire is blue topaz, and Brazilian ruby is either red tourmaline or yellow topaz, changed to pink by artificial heat; Scotch or "false topaz," is simply yellow quartz. Limpid pebbles consisting of genuine topaz are called *gouttes d'eau.* Mawe says the river-beds of Brazil yield white, blue, and sea-green varieties, and Burton found both the topaz and the ruby in the itacolumite of this country; the former has been known to occur in granite, and possibly it may exist in other kinds of rocks.

Topaz is seldom found in large crystals without defects, but it sometimes constitutes massive rocks which, in Saxony, are called "topaz fels." Specimens of immense size in a crystalline form have been taken from the Urals; one of this description in the collection of St. Petersburg, of a wine color and perfectly transparent, has been differently estimated to weigh twenty-two and one-half, and thirty-one pounds; another, found in Scotland, has a weight of nineteen ounces. A topaz described by Tavernier, belonging to the Emperor Aurungzeeb, which was purchased at Goa for nearly sixty thousand dollars, weighed one hundred fifty-seven and three-fourths carats. Very fine specimens of this gem were exhibited at the London Exposition of 1851, from New South Wales and other regions; the brilliant mineral sent from Russia as phenakite, when subjected to the test of specific gravity, proved to be topaz.

It has been said that this precious stone has never been engraved, but this is an error, on good authority, since several of this class are known to exist. An antique engraving bearing a star or a cluster of stars has been counted with the treasures of St. Petersburg; another, engraved with the portraits of Philip II. and Don Carlos, is in the Royal Library of Paris; and a third, inscribed with a motto in Arabic, is thought to be owned in the same city. As an ornamental stone, the topaz is less popular than formerly; therefore, its commercial value is small compared with some other gems.

CHAPTER XV.

OPAL.—PEARL.

THE name opal, it is thought, was primarily derived from the Sanskrit *upala*, meaning a stone or rock, more directly from the Greek and Latin *hopallios*, *opalus*.

Of all precious stones, says Pliny, the opal is the most difficult to describe, since it seems to combine in one gem the beauties of many other species,—the fire of the carbuncle, the purple of the amethyst, the green of the emerald, and the yellow of the topaz.

Many speculations have been advanced by chemists in regard to the causes of the remarkable peculiarities of this precious substance, some maintaining the opinion that its beautiful play of colors depends principally upon the quantity of water it contains, which varies greatly in different varieties, while others believe that water is not absolutely essential to produce this striking effect.

The brilliancy of the tints is heightened by heat, unless too intense or too prolonged, when the colors vanish entirely. This result seems to indicate that the presence or quantity of water has some connection with the development of the wonderful iridescence of the opal.

It may be of some interest to know that so distinguished a philosopher as Sir Isaac Newton believed the play of colors in this gem were the result of the refraction and reflection of light caused by the exceedingly small and numerous fissures crossing it in every direction.

Newton's theory has its objectors, some of whom argue that the beautiful, variegated reflections which lend such a charm to this gem are caused by laminæ, a structure found in some other minerals. Whatever may have produced the exquisite tints, it is conceded that its special attraction is due to its imperfections, and not to any coloring matter, as in the case of most other precious stones.

In the limpid varieties, it has been stated, it is difficult to perceive any cause for the play of colors seen in this gem, but in the translucent specimens the case is entirely different. Thin films or clouds are observed in the interior, of a slightly reddish tint, floating, as it were, beneath the surface, but which change their color as the stone is turned so as to receive the light at a different angle. Films, apparently alike, often exhibit different phenomena, some presenting but one color, some two, while others display a continuous spectrum, and not unfrequently this beautiful play of colors is completely eliminated by polishing. Sulphuric acid causes the opal to turn black, an effect which has led to the opinion that it contains organic matter.

The composition of this mineral is not fully assured. Silica forms its principal constituent, as in quartz, but in a different condition, being amorphous, soluble, and usually hydrous, though water is not considered essential. Its specific gravity and hardness are less than those of quartz, and it is generally supposed to be incapable of crystallization.

Opal has been found in the cavities and fissures of volcanic igneous rocks, porcelain earth, limestone, and the silicious waters of hot springs. It is a natural production of Ceylon, Australia, Hungary, Iceland, the Hebrides, Ireland, Faröe Islands, Honduras, Mexico, and the United States, though here it is not usually in a condition suitable for gems. Pliny, who knew nothing of the Western continent, says: "India is the sole parent of the opal, thus completing her glory as being the great producer of the most costly gems," but he admits that either a variety of this species, or one closely allied to it, denominated "lovely youth," was brought from Egypt, Arabia, and some other regions of the East.

The principal varieties are the precious or noble opal, common opal, fire-opal, jasper-opal, wood-opal, girasol, cachelong, hyalite, hydrophane, asteria, and a kind exhibiting dendritic markings, sometimes called moss-opal. When the colors are broken into small masses, it passes under the name of harlequin-opal; and when characterized by an orange hue, it is golden opal.

The noble or precious opal, giving out different colored rays in bewildering succession, constitutes one of the most beautiful ornamental stones in existence, and has always been regarded as one of the most desirable and attractive for personal use. When employed in jewelry, it is cut with convex surfaces, *en cabochon*, on both sides. Camei are sometimes carved on this gem, in a manner to present the figure on a ground consisting of the dark brown matrix.

The fire-opal, found in Mexico, Hungary, and the Faröe Islands, is characterized by its remarkable flame-like reflections of hyacinthine red, passing to honey-yellow, and sometimes presenting all the prismatic colors. The two largest fire-opals known in England, according to Streeter, were found in the Hungarian mines in 1866, and exhibited at the Paris Exposition of 1867. They are drop or pear-shaped, one of the gems weighing one hundred and eighty-six carats, and the other one hundred and sixty.

The variety known as *hydrophane* was so named from its peculiar property of becoming transparent when plunged into water. In its ordinary state it is white or reddish yellow, slightly translucent or completely opaque; but when immersed in water, it emits bubbles of air and changes its appearance sometimes exhibiting the prismatic colors of the precious opal, losing them, however, on being removed from the water. The hydrophane is developed in Saxony, Hungary, France, and Italy.

The opal which figures in "Anne of Geierstein" is represented as suffering a loss of beauty by contact with water; therefore, after the publication of the novel, this beautiful stone acquired the reputation of being an unlucky gem, and was for a time discarded by fashionable circles, but it has since been restored to its legitimate rank, though there are persons who still cherish the opinion that it is an ornament of ill omen.

The *cachelong*, a name signifying "beautiful stone," found in the River Cach, in Bukhara is nearly opaque, with whitish, yellowish, or reddish colors. The *girasol*, meaning "to turn to the sun," is a translucent variety, wearing a bluish white color, which gives out red reflections in a bright light; the fire-opal, when of a hyacinth-red, is sometimes called girasol or sun-opal.

Trees and other vegetable products are not unfrequently silicified or petrified by opal, constituting a variety called *wood-opal;* but they do not display the prismatic hues of the true opal. Another form of this chameleon-like gem is afforded by different shades of color arranged similar to those of the agate, and is called *opal-agate.* The *jasper-opal*, unlike most other varieties, comprises some foreign substances which give it the color of yellow jasper, though it retains the lustre of the opal. *Hyalite* is a colorless, transparent kind, sometimes called "Muller's glass"; the *star-opal* gives out sudden flashes of color like lightning from the clouds; *common opal* is a translucent, non-prismatic variety, of a milky-white tint inclining to green, yellow, or blue, and is used in Germany for cheap jewelry.

The opal was prized by the ancients above most other precious stones. The Romans obtained their supply from the East, perhaps from Ceylon, where they are now found, but the largest did not exceed the size of half a hazel-nut, except the famous opal of Nonius, who preferred exile rather than surrender it to Mark Antony. This historical gem has been variously priced from one hundred thousand to nearly one million dollars,— quite a difference to reconcile, but it proves the high esteem in which the opal was held at that time.

This gem is very difficult to engrave, and sometimes quite impossible, yet there are a few antique works of the kind, including one belonging to the Orleans collection, and another in Paris engraved with the portrait of Louis XIII. Some rude intagli, apparently antique, are occasionally found in this stone, usually of the opaque varieties. There is a fine specimen in the Praun collection, engraved with the heads of Jupiter, Apollo, and Diana.

On account of its softness, frangibility, and liability to injury from oily substances, it is not suitable for ring-stones, but may be used, with proper care, for other ornaments. Unfortunately, this desirable gem is affected by atmospheric influences, as severe cold, which, it is thought, causes exterior flaws tending to extinguish all its "fire," and reduce it to a common pebble. The only essential remedy for this defect is to remove the outer layer, but this operation is open to objections since it diminishes the thickness of the stone, and allows a freer passage for the light through it, and, as a result, its beautiful iridescence is impaired or lost.

The opal was counterfeited more successfully by the ancients than any other gem, so that it was nearly impossible, with their tests, to distinguish between the real specimens and their imitations, but the knowledge of this art has been lost, and modern attempts to revive it have not been satisfactory.

This mineral exists in nature from transparent to opaque,—that from Queensland and Honduras affords specimens as clear as glass—yet an intermediate state of this quality, if the play of colors is well developed, is the most highly prized. There is no fixed standard of value for this commodity, as much depends upon the "fire" exhibited in different stones of the same size, and the price may range from five dollars to five thousand. Black varieties, which are rare, bring a great price at the present day, though the sum of several thousand dollars paid for one of this color is a trifle compared to what was given in the Roman Empire for other varieties.

The mines of Hungary, and those of Honduras and Mexico, are the most celebrated of modern times. Those of Hungary, discovered in the fifteenth century, are situated on a mountain branching off from the Carpathian range near the village of Czerniska, or, it may be, Czernowitz, on the River Pruth, in Galicia; the quarries are skilfully worked and yield the hardest and most durable opals known. It is stated that the gem, when first taken from the mine, is transparent and colorless, but after exposure to heat and light, it contracts in size and begins to display its natural iridescence, which increases in variety and splendor until the excess of moisture is expelled. Mr. Hamlin mentions, in connection with this subject, that solar heat produces finer colors than artificial, and that the violet hue invariably appears first.

The Honduras deposits are thought to be extensive, but they have been only partially explored and imperfectly worked. The opals are found in porcelain earth, in small, irregular masses, or of globular shape, and pale tints, from brown to pearl gray, though not unfrequently they exhibit rich prismatic colors. They are softer than the Hungarian variety, but compare favorably with it for brilliancy and durability. It has been reported that one of the richest opal-mines in the world has, within a few years, been opened in the province of Queretaro, Mexico, yielding numerous fine specimens with a great diversity of colors, including blue, pink, red, green, yellow, cream color, and black. The fire-opal, the most resplendent of all the different kinds of this wonderful gem, is found in the greatest perfection in porphyry, at Zimapan. It is translucent and emits brilliant fiery-red, yellow, and green reflections; but it is easily impaired by exposure to moisture and changes in temperature.

Remarkable specimens are known to occur in different collections. There is one in the imperial cabinet of Vienna, found at Czernowitz, near the Pruth, in 1770, which weighs seventeen ounces, and, not withstanding its cracks and fissures, the sum of fifty thousand dollars has been offered for it, but the government refused to sell it, even at that price. The finest Hungarian opals are seen among the crown jewels of Austria, though France numbers among her state collections two very valuable gems of this kind. Probably the most remarkable opal on record was the one owned by the Empress Josephine, which was called the "Burning of Troy," on account of the innumerable red flames it emitted, as if on fire. The under side was perfectly opaque, but the upper portion, being transparent, served the purpose of a window through which were seen the glowing rays of fiery light, very appropriately compared to the conflagration of a great city.

Magnificent examples of the opal have been frequently exhibited among the curiosities of modern expositions. A

Honduras specimen, weighing six hundred and two carats, valued at twenty-five thousand dollars, together with very beautiful green and purple varieties from Queensland, and fire-opals from Mexico, attracted the notice of visitors to the Centennial at Philadelphia, and the collection at the Exposition in New Orleans embraced a great variety, including jasper-opal, cachelong, hyalite, black opal from Bohemia, semi-opal of a snuff-brown color, precious opal, wax-opal of dark variegated tints, from Hungary, a variety from Tripoli, resembling a reddish sandstone in appearance, black wood-opal from California, brown wood-opal from New Zealand, a kind named geyserite, from Yellowstone Park, yellowish, light brown, and white from Colorado, and opals from the Island of Elba. It is related by Mr. Hamlin that one of the most beautiful jewels seen in this country is a necklace made of opals obtained from Honduras, cut and mounted in gold,—with diamonds. They were secured by Dr. J. Le Conte who has given some important facts about the mines of this region, in his report of the Inter-Ocean Railroad Survey.

The Pearl.—This precious substance has been considered from time immemorial one of the most beautiful and valuable productions of earth, and has been sought, at almost infinite labor and expense, as one of the loveliest gems that ever graced a coronet.

Pearls, in the strictest sense, cannot be called precious stones, since they have their origin in the animal kingdom; they are thought to be concretions of carbonate of lime and organic matter found in certain animal species, deposited in thin films which overlie one another, thus causing the beautiful iridescence characteristic of these organic productions. They are found both in marine and fresh-water mollusks, usually the pearl-oyster and the Unio, a fresh-water mussel, though nearly all bivalves with nacreous shells occasionally yield pearls. Those from Ceylon, the Persian Gulf, Madagascar, Australia, Panama, and California, are derived from the pearl-oyster; while those of Scotland, England, and Wales are obtained from the Unio, and are generally inferior to the oyster pearl in the iridescent sheen called "orient," though specimens of great beauty are occasionally discovered in the mussel. A pearl of great purity from the Conway River, in Wales, forms a conspicuous gem in the royal crown of England.

Nature has furnished this exquisite jewel with a charming and convenient casket lined with nacre, or mother-of-pearl, smooth, lustrous, and opalescent, where the little creatures dwell that afford the pearl, which does not apparently constitute a part of the animal but seems to be something foreign to it. How did it get into the shell, and why is it found in some shells of the same species and not in others? Different opinions are afloat about the origin of pearls, some theorists attributing them to accident, others to purely natural agencies. The ancients had a pretty correct idea of their origin; Theophrastus thought they were the product of a kind of oyster, the Pinna marina, found in the Indian and the Persian seas; but as to the manner of their formation, they differed in opinion, as their successors have done. Some of the writers, as Pliny and Dioscorides, believed they were caused by rain and dew falling into the shells of certain mollusks, an opinion alluded to by Moore in the following lines:—

> "And precious the tear as that rain from the sky
> Which turns into pearls as it falls in the sea."

Others have considered pearls the offspring of tears, therefore of suffering,—an idea which seems to foreshadow the modern belief that they are the result of injury. There have been those who thought they were propagated as animal species are, and others who ascribed their formation to some peculiar element in the water favorable to their production. The Hindoos refer the pearl to the creative power of Vishnu, and the Chinese have a tradition that a rainbow gradually descended to the earth in the form of an immense pearl. To make this theory complete, we must suppose the rainbow was broken into innumerable fragments, each perfect in itself, thus furnishing all the gems of this kind ever since. One of the most poetical though not the most scientific hypotheses, referred to by Streeter, is given by the Persian poet Saadi, in this wise: A drop of water, falling into the sea one day, became ashamed and confused at finding itself in this immeasurable expanse, and exclaimed, "What am I in comparison with this vast ocean? My existence is less than nothing in this boundless abyss." While in this mood, a shell received the modest drop of water, and it became a magnificent pearl, worthy to adorn the diadem of a king.

Modern scientists, as has been intimated, do not agree about the cause of those rounded concretions called pearls, found in the shells of some bivalves; the most common opinion among them has been that they were the consequence of some disease of the animal, originating in a foreign substance introduced into the shell, as a grain of sand, or a parasite, and to relieve itself of the irritation naturally resulting from such a cause, the creature enclosed the intrusive body with pearly secretions, and a beautiful gem was produced. This supposed origin has sometimes furnished an illustration of the adage that the most noble achievements have been the result of painful efforts.

There is, probably, some truth in this hypothesis, remarks Mr. Streeter, but microscopic examination proves this was not the only, nor the necessary method of their production, since some pearls are empty while others are solid and of the same texture, color, and formation throughout. Another view is that the pearl is formed from the superabundance of calcareous matter designed for building up the shell, as they are similar in composition and appearance.

The *exact* nature of the secretion has not been satisfactorily ascertained, but it is supposed to be carbonate of lime and animal tissue, consisting of concentric layers secreted by the "mantle" of the mollusk in the same way the shell is secreted, except in the latter the layers are parallel. Linnæus, who believed the pearl to be the result of injury to the oyster, conceived the idea of introducing some foreign matter into the shell, and thus obtain the genuine article by a forcing process. The East Indians and Chinese adopted a similar method, and secured a product similar though inferior to the natural gem.

The lustre of the pearl is peculiar to itself, and it has been said, has never been perfectly imitated, but, unfortunately, they lose their beautiful reflections by age, acids, gas, and other noxious vapors, and may in time crumble into dust; to prevent their loss of brilliancy, they should be kept in dry magnesia. Deteriorated pearls may be restored to their original beauty unless the injury penetrates to the centre. Those set in antique jewels have rarely been found uninjured by the effect of age or other agencies.

Round pearls are most admired, but oval or pear-shaped ones are much larger; the oriental specimens are nearly always round and of a white or yellowish tint, while those from Panama are generally drop-shaped and of a dark color. Misshapen pearls, called "baroques," are frequently met with as curiosities; they were employed by the Cinque-cento jewellers for grotesque pendants worn as ornaments, but at the present day these "freaks of nature" are to be seen only in collections and museums. One of the largest groups of these "baroques" is found in the Green Vaults, at Dresden.

The Devonshire collection includes a very large pearl of this kind, which personates a mermaid, and is valued at ten thousand dollars; another of these abnormal productions, presented to the Great Mogul by Tavernier, as a gift from his government, represented a siren arranging her hair.

Pearls are of various colors, comprising white, black, rose, salmon, blue, gray, and pink, which are sometimes imitated by pink coral; those found in the Western continent are of several different shades. Black pearls command a high price at the present time, on account of their rarity, but in Tavernier's day they were of little account, especially in the East. "The orientals," he says, "prefer the whitest pearls and the blackest diamonds." The Persians arranged this gem in twelve different classes, according to its form and color.

A pearl of the first quality must be iridescent, of bright lustre, and pure whiteness, or of a delicate azure tint, which is the most highly esteemed, those of a yellowish hue being considered of inferior quality. When used in ordinary jewelry they are cut in halves, or perforated for beads, an operation requiring great care to prevent their splitting. Their commercial value, like that of many of the precious stones, has always been fluctuating, according to the changes of fashion or the fancy of collectors. The price at the present day depends largely upon their form, color, texture, and "water." Some famous pearls on record were estimated at fabulous prices; that of Cleopatra was valued at one million sesterces, or perhaps four hundred thousand dollars; that of Sir Thomas Gresham, at seventy-five thousand dollars. Their use as a personal ornament has been equally vacillating, they having been advanced at one period to the first rank among gems, and at another, consigned to oblivion. In France, they reached their climax during the regency of Catherine de Medici, where they were preferred to any other gem, until they were superseded by the diamond, in the time of Louis XIV.

The largest pearl known to Pliny weighed a little more than fifty-eight carats,—a magnitude not often equalled, except by baroques ; the finest in the French regalia, as shown by the inventory, did not exceed twenty-seven carats. The celebrated pearl of Phillip II., known as "La Pelegrina," belonging to the Spanish crown, weighs thirty-four carats. It is pear-shaped and was obtained from Panama, or from San Margarita, on the coast of South America, and is valued at fourteen thousand four hundred ducats. The name "La Pelegrina" has been given to another pearl, claiming to be the largest perfect specimen known in Europe, weighing one hundred and twenty carats. It was brought from India and sold to Philip IV., King of Spain, but is now, it is said, in the possession of one of the noble families of Russia. A pearl owned by one of the sovereigns of Persia weighed one hundred and sixty-eight carats, and was valued at two hundred and eighty thousand dollars. The present Shah of Persia and the Imaum of Muscat each own a gem of this kind, of inestimable value. A pearl remarkable for beauty, transparency, and perfection of form, though of little more than twelve carats in weight, was owned by an Arabian prince, who refused to sell it to the Moslem emperor Aurungzeeb for one hundred and forty thousand livres. The specimen in the Hope collection, South Kensington, considered one of the largest known, is pear-shaped, and yields the remarkable bulk of three ounces.

Tavernier mentions several pearls distinguished for size, one of which, owned by the Great Mogul, served as pendant to his chain of emeralds and smaller pearls, and another was suspended from the neck of the peacock adorning the famous throne. Among the treasures acquired by Saladin during his conquests were seven hundred pearls of priceless value. The crown jewels of France, in 1791, included fifty-three pearls, valued at that time, altogether, at nearly one hundred and fifty-three thousand dollars.

The finest pearl necklaces of modern times are said to belong to the royal families of England and Germany, and to Eugénie, late Empress of France; the English jewel was the gift of the East India Company to her Majesty Queen Victoria. An interesting domestic incident is related of the imperial family of Germany, how that the empress received from the emperor a single pearl on her first birthday after their marriage, to which he added a new one on each succeeding anniversary, until a magnificent necklace has been formed. The Crown Princess of Germany has a necklace composed of thirty-two pearls, valued at nearly one hundred thousand dollars.

The statement that the pearl has been dissolved and drunk is equivocal. The story of the famous pearl belonging to Cleopatra is well known, but its romance is spoiled by some investigating chemist, who has decided that there is no acid safe to be taken into the stomach which could dissolve this substance. Others accept the tradition as a fact not to be questioned, since an acid such as vinegar, they say, can dissolve it, provided the experiment is sufficiently prolonged On the testimony of Pliny, the practice of dissolving pearls for a beverage was known before Cleopatra's day; the liquor thus formed possessed a delicious flavor, and it is said to have constituted a favorite drink of the Emperor Claudius, at a later period. A similar feat, rivalling in folly that of the Egyptian queen, was performed by Sir Thomas Gresham, an English merchant, in the reign of Elizabeth. When the Spanish ambassador, as the story goes, was extolling the riches of his sovereign, in the presence of the queen, Sir Thomas replied that her majesty had subjects who at one meal expended a sum equal to the daily revenues of the King of Spain and all his grandees put together.. Soon after this interview, the ambassador was invited to dine with the English knight, when the latter drew from his pocket a pearl for which he had refused seventy-five thousand dollars, then ground it to powder and drank it in a glass of wine to the health of the queen. This incident does not, however, settle the question whether the pearl can be dissolved by a harmless acid, since the Englishman's gem was pulverized, and drunk as a powder.

There is no doubt of the great antiquity of the pearl used as a gem, nor of the high estimation in which it was held, as it is mentioned by early writers, both sacred and profane. Homer, who is unaccountably reticent about precious stones, is thought to refer to pearls in the term "triple-eyed," applied to Juno's famous necklace,—an interpretation supported by the fact that a triplet of pear-shaped pearls forms a distinctive feature in the antique heads of this goddess. The Persians cherished the greatest admiration for this ornament, as is shown by the custom of the Sassanian kings, who are always represented with an enormous pearl in the right ear. The Greeks imitated the Persians in the use of this gem for personal decoration, and the Romans surpassed both nations in their extravagance, often paying exorbitant prices for it, regarding the possession of a single costly pearl of more importance than the conquest of a province. In one of his triumphal processions into Rome, Pompey displayed thirty-three crowns, many of them taken from the treasury of Mithridates, a portrait of this prince, his own bust, and a small temple of Mars, all made of pearls. Julius Cæsar found them very abundant in Great Britain, and the Romans, during their occupation, gathered great quantities of them from the fresh-water mussel. A breastplate made of these British pearls was dedicated by the great conqueror to Venus Genetrix, at Rome.

These beautiful gems have served the purposes of metaphors and other rhetorical figures in literature. They were employed by the sacred authors as emblems of whatever is superexcellent and difficult to obtain. They entered into the list of precious materials used in the construction of the celestial city, whose twelve gates were made of pearl. Other writers, as Shakespeare, Milton, Moore, and Scott, have selected this gem for rhetorical effect. The Arabs compared eloquence, which they classed with the most important accomplishments, to pearls. Babylonians, Persians, Egyptians, Greeks, and Romans consecrated them to their divinities, an indication that they valued them as the choicest productions of earth. They have been esteemed for their supposed medicinal and mystical virtues. Marco Polo informs us that the King of Malabar wore a kind of reliquary of rubies, sapphires, and emeralds, with a pendant of one hundred and four large pearls and rubies, by which he counted his prayers to his idols, morning and evening.

Pearls have been obtained from certain localities in both hemispheres. The oriental fisheries include those of the islands in the Persian Gulf, where they are sometimes taken at the surprising depth of twenty fathoms; the Red Sea, formerly the chief source of supply, but now nearly exhausted; the Indian Ocean, Ceylon, the East India Islands, and Japan. There have been, sometimes, thirty thousand natives, yearly engaged in the business of pearl-fishing in the Persian Gulf, at a profit of more than three hundred thousand pounds per annum: the most beautiful specimens come from this region. The coasts of California and British Columbia, and the Gulfs of Mexico and Panama, are the chief places for pearl-fisheries on this continent. These localities were, probably, known to the Aztecs, since the Spaniards found immense quantities of pearls among their treasures. Cortez received from Montezuma, whose palace was studded with pearls and emeralds, rich presents, consisting of pearls, while a certain Mexican chief proffered an annual tribute of one hundred pounds of these gems to the King of Spain. Soon after the conquest, the Spaniards established fisheries on the South American coast, which caused so large an importation into Europe that, according to one of their historians, they were sold in heaps at public auction in Seville. These fisheries were exceedingly productive when first opened, and for a long time subsequently; but the supply, after a while, gave indications of exhaustion, though specimens of fine lustre but irregular shape are still gathered on the west coast of Central America. The pearls of Columbia and California, said to display a peculiarly beautiful lustre, have been considered the peers of the Panama varieties. Some of considerable size have been found in New Jersey. A pearl from this region, discovered several years ago, exceeded an inch in diameter, and was sold in Paris for more than two thousand dollars, — a paltry sum, it may be thought, when compared with the value of some of the celebrated pearls of history.

Both the Scotch and the English varieties acquired an early reputation and were thought to rival the oriental specimens. It is estimated that the fisheries of the Tay and the Isla, during a period between 1761 and 1764, amounted to fifty thousand dollars; but subsequently the yield declined until 1864, when it revived, and in that year sixty thousand dollars were the results of the production. The rose-colored Scotch pearls, obtained from the fresh-water mussel, frequently found of considerable size, sometimes weighing several carats, are regarded with general favor in England. Pearls are also found in Ireland, Bohemia, Germany, France, Sweden, and Russia. The European varieties, even when large and well shaped, are inferior in brilliancy and play of colors to the oriental and the American pearls. The best specimens occur in the oldest shells, considerable time being required to bring them to perfection.

They are sometimes obtained by dredging, but usually they are brought up from the rocks in the ocean, to which they attach themselves, by divers trained for the purpose. The business is attended with danger, largely from sharks, but in regard to its effect upon the health of the divers, different views are taken, some judges considering it injurious, while others do not regard the occupation as especially harmful to the constitution. The use of the diving-belt, — a modern contrivance, — will, it is believed, greatly reduce the perils to which divers are exposed. The natives trust for safety in the power of the shark-charmers, who pretend to render these dangerous monsters harmless by certain incantations; at the English fisheries in Ceylon, they are paid by the government, whose policy it is to keep up the delusion.

The time during which divers can remain under water varies from about two to six minutes, which may be extended by modern appliances, and experts make from forty to fifty descents in one day, securing, on an average, one hundred oysters each time. The greatest depth reached by these fishermen is reckoned to be fifteen fathoms, — ninety feet, — while the usual depth is nine fathoms, or fifty-four feet. Their descent is accelerated by artificial weight, each diver carrying about twenty-five pounds of rocks for that purpose.

The oysters when secured are left in heaps to die before the shells are opened, and they are frequently sold in this condition, a practice which gives the business something of the character of a gambling operation. Some shells yield no pearls, while others contain from one to one hundred or more, of different sizes, from seed-pearls to those of ordinary and even unusual dimensions.

The mother-of-pearl, or lining of the oyster shell, constitutes no small item in the profits of the business. It is computed that fifteen thousand tons of this commodity, the product of five or six millions of oysters, are annually imported into Europe. The pearl, when first detached from the shell, exhibits a slight roughness at the point of contact, which is removed by polishing with pearl-powder.

The Ceylon fisheries for the first two years after the English occupation of the island, in 1797, yielded pearls amounting to nearly two million dollars; at present, they are closed to afford time for the oyster to recruit. A description of these fisheries is given by an eye-witness, who says the "season" at Ceylon comprises from four to six weeks in March and April, and presents a busy scene. The shores of the Bay of Candalchy, the place of operations, afford a perfect babel of tongues; here are people representing many different nationalities, with all the appliances of the business, commingled; with tents, huts, and markets thronged with jewellers, merchants, and traders, all in eager pursuit of the same object, — the acquisition of wealth. The number of oysters obtained during the season is amazing; a single boat has been known to land in one day thirty-three thousand oysters, which, augmented by the number fished up by all the other boats, exceeds all computation. This business holds out a tempting bait for the dishonest employee, and, notwithstanding the utmost vigilance, thefts are frequently perpetrated by all classes, divers, boatmen, washers, sifters, and even superintendents, who have been known to extract the pearls from the washing-troughs by attaching a viscous substance to the end of the canes used for punishing delinquents for the same offence.

The most common varieties are sent to China, those of the next higher quality are exported to Poland, South Germany, Russia, and the Danubian provinces, where they are worn by the peasantry. Oriental princes have been the readiest purchasers of the finest South American pearls; Goa, in India, was once the greatest mart in Asia for pearls, as well as for diamonds, rubies, sapphires, and other valuable gems.

Artificial pearls are made of small globes of glass lined with wax and scales taken from the living fish, so as to preserve the glistening hue.* Roman pearls differ from other imitations in having the coating on the outside of the glass. A variety of the smelt, inhabiting the Tiber, affords the Roman jeweller with the means of making wax beads more closely resembling the genuine pearl than do either the Venetian or the French counterfeits.

* A method somewhat different from this consists in putting into the hollow glass bulbs a mixture of liquid ammonia and the white substance of the scales of certain fishes.

CHAPTER XVI.

SPINEL, GARNET, TOURMALINE, TURQUOISE, LAPIS-LAZULI, ZIRCON, CHRYSOLITE, CHRYSOBERYL, IOLITE, KYANITE, APOPHYLLITE.

Spinel.—The name spinel is said to mean "spark," and is so called, probably, from its pointed crystals in the form of octahedrons. It has frequently passed for oriental ruby, but it differs from that gem in its chemical nature, having for its constituents alumina and magnesia with traces of certain oxides in the colored varieties.

This precious stone affords a wider range of color than almost any other, including all the prismatic hues with their different shades and combinations, besides the colorless and the black varieties; crystals occur from perfectly transparent to nearly opaque. The kinds used for jewelry are *spinel-ruby*, of pure red or crimson, tinged with blue or brown; *balas-ruby*, exhibiting a ruby-red diluted with rose or lilac; *rubicelle*, yellow or orange-red; *almandine*, of a violet hue; and *Ceylonite*, or *pleonast*, green and dark brown to black, steel-gray, or slate. All these colors afford numerous gradations in shades.

Before its composition was understood, there was no distinction made between the spinel and the corundum ruby, which accounts for the fact that so many of the celebrated rubies, so regarded, have proved to be what are called by modern mineralogists spinels. De Lisle, in 1783, was the first scientist to distinguish between these different gems. Both the spinel and the balas receive the name of ruby among jewellers, yet their commercial value is far less than that of the true ruby. The spinel is deficient in the prismatic play of colors owing to its small refractive and dispersive powers, but, aside from this imperfection, it rivals the corundum gems as an ornamental stone.

Various opinions have been given about the origin of the name balas; Marco Polo thought it was derived from Balla-heia, a mountain in India. Chardin believed it came from Baluchani, a place in Pegu; hence it is called "the stone of Balachan," the Persian name for ruby. King traces its origin to Balashan, in the neighborhood of Samercand, where it is found; while another writer says, with some hesitation, that the name balas, or balais, is a derivative of Beloochistan, which was formerly called Balastan, where the mineral was discovered in the thirteenth century. The term almandine applied to a variety of spinel of the hue of the almond blossom, is simply an epithet to designate color, and is also given to a variety of the garnet.

Spinels suitable for jewelry are found in many different countries — Burmah, Siam, Ceylon, Sweden, Bohemia, Austria, and the United States. A few very good gems, of small size, have been found in California, and specimens of smoky blue, green, and dark claret, weighing not more than two carats, have been discovered in New Jersey, while Sweden yields a blue variety, and Mount Vesuvius a black; but the finest and largest spinels, occasionally weighing from twenty-five to one hundred carats, are obtained from India. Mr. Streeter mentions two stones of Indian origin imported into England in 1861, which weighed, respectively, one hundred and two and one-fourth carats, and one hundred and ninety-seven carats. De Berquem refers to a table balas belonging to the Shah of Persia, which weighed two ounces. Fine specimens occur as pebbles in the beds and on the banks of the rivers of oriental countries; but according to a Persian tradition, the mines of spinel were revealed by the opening of a hill at Chatlan, during an earthquake.

One of the finest spinels known, and equal in size to a pigeon's egg, is in the possession of the King of Oude. Tavernier enumerates one hundred and eight large rubies in the decorations of the throne of an Indian monarch, varying in weight from one hundred to two hundred carats, while the computed size of one of them was two and one-half ounces. These gems are now supposed to have been balas-rubies, and are placed in the same predicament with the famous spinel in the English crown, once thought to be a ruby, a gem of historical interest, having been owned by Don Pedro of Castile, then by the Black Prince, and afterwards worn by Henry V., at the battle of Agincourt.

The Garnet.—The garnet group of minerals includes several species, having no other characteristics in common than their chemical composition and form of crystals, while in color, hardness, and specific gravity they differ very essentially. The crystals of the garnet are cubical, singly refracting, monochromatic, and transparent to translucent; in hardness, this gem-mineral ranges from a little below to a little above quartz, and includes different reds, yellow, green, brown, black, and white. The term is derived either from pomegranate or granatus, like a grain, and it is supposed to be one of the precious stones to which the name anthrax was applied by Theophrastus, and carbunculus or Alabandine stone, by Pliny.

It is not a rare gem, but occurs in many localities in both hemispheres; the best American garnets are found in Colorado, New Mexico, and Arizona, which are said to yield several thousand dollars worth of gems annually, but with a capacity for a much larger production. These garnets, including blood-red, almandine, yellow, and other colors, are thought to be as fine as those from any other country.

The numerous varieties of the garnet are named according to the color of the mineral, its native home, or some other casual circumstance, and comprise: *almandine*, or precious garnet; *essonite*, or cinnamon-stone; *vermeille*, or hyacinth-garnet; *succinite*, an amber-colored variety from Piedmont; *pyrope*, or Bohemian garnet; *grossularite*, from Siberia, of a pale green; and *uwarowite*, from the Urals, of a beautiful emerald green and remarkable brilliancy, but seldom of sufficient size and transparency for gem-stones. The Italians give the name *jacinta la bella* to a yellow-garnet, *guarnaccino* to a yellowish crimson, and *rubino-di-rocca* to a variety tinged with violet. *Syrian*, or *seriam*, garnet is obtained from Syriam, in Pegu, and not from Syria, as is sometimes stated. A honey-yellow occurs in the Island of Elba, and a black variety, called *melanite*, is known in Italy and some other places. In fact, this precious stone assumes so many forms, it has very appropriately been called the Proteus of the gem family.

The *almandine*, or *almandite*, found in Ceylon, Brazil, Greenland, and other countries, one of the most beautiful of the species, is noted for its cherry, blood-red, or brownish tints, which assume an orange hue by candle-light, and is sometimes sold for rubies. The *Bohemian*, or *pyrope*, meaning "like fire," a native of Bohemia, Mexico, and South Africa, is a deep, clear red garnet and the hardest of all the varieties, ranking seven and one-half in the scale; some mineralogists make a distinction between the pyrope and the Bohemian. The *essonite*, or cinnamon-stone, presenting a gold color tinged with flame red, has often passed for hyacinth, a variety of the zircon of the same hue. The best specimens of essonite are imported from

Ceylon, a locality which undoubtedly furnished the ancients with this gem, since numerous antique intagli are found on this variety of the garnet. The dark orange *hyacinth* garnet is also sometimes taken for the true hyacinth, or red zircon.

The name jacinth, or hyacinth, is given to varieties of several species, as the garnet, the sapphire, the zircon, the topaz, and the Vesuvianite, and, like some other names, is only an epithet conferred on account of the color. Some lapidaries identify the hyacinth with essonite, and others regard it as distinct from the garnet, but its crystalline form and typical composition are identical with those of this species, the difference consisting in color and specific gravity with thirty per cent of lime in place of protoxide of iron. Engraved gems of what was thought to be true hyacinth are in reality either hyacinth garnet or sard.

Guarnaccino, the brownish red variety of the Italians, unites the qualities of the garnet and the spinel, and when of superior excellence, it can hardly be distinguished from spinel-ruby, while a rose-colored garnet resembles the balas-ruby. An orange-red variety receives the name *vermeille;* the star-garnet, which displays a star, or rather a cross, when held in the sunlight, owes this distinction rather to its construction than to its color. A beautiful gem of different greens shading to liver-brown, thought to be garnet, has recently been discovered at Bobrowska, Siberia, in nodular masses, from the size of a pea to that of a chestnut. It is a soft mineral, not exceeding five in the scale, but has a remarkable play of colors; its exact chemical composition is not placed beyond doubt. Beautiful white garnets, yielding gem-stones, are developed in Canada, and a coarse, granular variety, called *colophonite*, is found in Scandinavia and America.

The name *carbuncle*, as applied to a precious stone, is very bewildering, sometimes denoting the manner of cutting, and at other times a variety of several species. The ancients gave this term to all red stones in general, while modern writers are not much more definite in their application of the word. Theophrastus says it resembles burning coal, and emits light in the dark, is scarce and found only in few places, as Carthage, Massalia (Marseilles), Egypt, and some other localities. The Hebrew for carbuncle is a word meaning "lightning," and, according to a legend among the Jews, this precious stone was suspended in Noah's Ark, to diffuse light. In modern jewelry, the term is applied to the scarlet, deep-red, and crimson garnets cut *en cabochon.*

The garnet has always been extensively used for an ornamental stone both in ancient and modern times; the Greeks and Romans showed their predilection for it in their numerous engravings, while the Celts and Anglo-Saxons employed it for jewelry, granulated, filagree, and enamel work.

Though the garnet was quite generally used by the ancients for engraving, yet there are few good antique gems of the kind, and those belong mostly to the Roman school, which produced some fine intagli cut in this stone, especially those engraved with imperial portraits. The celebrated Marlborough garnet, engraved with Sirius, is considered a masterpiece of the glyptic art. The variety called carbuncle was frequently employed for engraving, as is known by several beautiful specimens seen in Paris, Turin, Rome, and St. Petersburg. The magnificent Atalanta, on carbuncle, contained in the Berlin collection, is considered one of the finest of the Greek school. Portraits of the Sassanian kings frequently appear on this species of precious stones, implying it was a general favorite with the Persian lapidaries, but on account of its brittleness, and, therefore, difficulty of being worked, it is seldom used by modern engravers except for small camei. It has sometimes been taken for ruby, and it is supposed by some judges that Wallenstein's ruby and many others seen by Tavernier in Bohemia were garnets, instead of rubies or spinels. The best specimens of oriental garnets are obtained from Ceylon and Pegu, and of these the latter are preferred to the former; the best European garnets are those found in Bohemia.

The Tourmaline. — There are several antique gems mentioned by old writers which had some of the characteristics of the modern tourmaline, affording a pretext for identifying any one of them with this mineral. Theophrastus refers to a stone found at Cyprus, which was green at one end of the crystal and red at the other; and Pliny says of the lychnis that it attracts chaff and filaments of paper when heated by the sun or by friction. Both these qualities belong, though not exclusively, to the tourmaline.

The true nature of this substance was not understood until within a century, and even at present its peculiar qualities, which render it an interesting object of scientific study, are understood by only a few. It stands almost alone in the mineral kingdom, at least among precious stones, for its constituents and physical properties.

Though the remarkable characteristics of the tourmaline were made known to the French Academy of Sciences in the beginning of the eighteenth century, yet they did not attract general attention until several years afterwards, when, by the published accounts of a German experimenter, the interest of the scientists of Europe in this mineral was suddenly aroused. At first the subject excited opposition, and a "paper warfare" followed, with a good deal of noise from both parties, which reached this continent and enlisted Dr. Franklin in the discussion. The quarrel in Europe was regarded of so much importance, or amusement, that Hogarth introduced the subject into one of his paintings. At length the controversy was ended by experiment, — a method which, it is supposed, ought to have been applied at the beginning, — and the result proved satisfactory to all parties.

The earlier mineralogists denominated this species of stone *schörl*, — a name thought to be derived from a village in Germany, and applied to the gem by the miners, from its association with this place. The identity of the schörl with the tourmaline was discovered, first by Linnæus, and subsequently by De Lisle, after they had been regarded as separate species for two centuries. Transparent crystals, cut as gems, were first introduced into Europe by the Dutch about the first of the last century, and with them their Cingalese name — tourmaline. The term schörl is now applied to a black variety.

The chemical substances forming this mineral are numerous, comprising a dozen or more different elements, silica and alumina constituting the larger part, while the composition varies in the different kinds. It has a hardness about equal to that of quartz. Crystals assume the form of prisms, but often terminate in a different manner, the positive end having a greater number of facets than the negative end, a circumstance of rare occurrence in crystallography. It is transparent to opaque, but transparency is exhibited only in one direction, — a fact important to lapidaries.

Some curious phenomena are exhibited by this wonderful mineral, which seem to invest it with almost magical powers. When two slices, cut parallel with the axis, and laid one upon the other, are viewed in one direction they are both transparent, but when seen in another direction they are opaque. If a doubly refracting crystal is placed between the two plates, the part covered by the intervening crystal is transparent, while the remainder is opaque. The tourmaline possesses double refraction and remarkable electrical properties, with great power of polarizing light, which is possessed in different degrees by specimens from different localities. Those from Mount Mica, in Maine, have it less powerfully than those from Canada and Brazil. When heated or excited by friction, it acquires different degrees of electricity at the different extremities of the crystal, and if broken in this state, the fragments present opposite poles like the magnet, but when subjected to a heat of about 212° Fahr., it loses its electricity but regains it by reheating, though with poles reversed.

The electrical property of this precious stone was accidentally discovered, it is said, by some children at Amsterdam, while playing with specimens brought there to be cut. They noticed that after exposure to the sun the tourmalines attracted or repelled ashes, straws, or other light bodies with which they came in contact, and communicated their observations to their adult friends. This naturally led to an investigation, which resulted in establishing the fact that this gem possesses remarkable electrical powers which render it one of the most curious and interesting in the whole mineral kingdom. On account of its attracting ashes, the Dutch lapidaries styled it *Aschentrecker*—"ash-drawer." It affords also an instance of pleochroism in its greatest perfection. Several other precious stones exhibit various colors in the same crystal, but the combinations presented by the tourmaline are, in many instances, peculiar to it. Some specimens, when viewed in the direction parallel to the axes, are crimson, but when slightly turned from this position they are white, smoky green, or of some other hue. The colors are greatly diversified both in shades and combinations. Some crystals appear green at one end and red at the other; some are crimson tipped with black, some are yellowish mixed with carmine, some are dark green passing into indigo blue, others are parti-colored, and not unfrequently specimens are known to be red internally and green on the outside. Not all crystals possess this property of exhibiting different colors in the same degree, while some do not have it at all; its absence is most conspicuous in the light colored varieties, which also display double refraction to a less extent. The Island of Elba yields a variety of tourmaline whose crystals are red at one end, yellow in the middle, and black or brown at the other extremity; violet and brown colors are sometimes combined in examples from Ceylon and Pegu. The line of demarcation is well defined in some specimens, while in others the different colors imperceptibly pass into one another.

The best known varieties of the tourmaline are: *rubellite*, of a red or pink color; *indicolite*, blue; *schörl*, or *aphrizite*, black; and *achorite*, colorless. The red tourmaline of commerce is called Brazilian ruby; the green, Brazilian emerald, peridot of Brazil, or chrysolite; and the blue variety, the Brazilian sapphire. The colorless tourmalines, found in Siberia, St. Gothard, and Elba, are very rare; those from Brazil, comprising nearly every shade of green and blue, yield some fine specimens. One of the most beautiful groups obtained from the Brazilian mines, and owned by a Grand Duke of Florence, exhibited a cluster of four erect crystals and one prostrate, all measuring from two to four inches in length, arranged in a matrix of one square foot.

This mineral, with the exception of the black tourmaline of Ceylon, is enclosed in gravel beds, rarely in nodules, and in granitic rocks as in the case of the remarkable beds of Maine. It is not rare as a species, having been discovered in many different countries of the globe. Brazil yields a large proportion of the specimens used for ornaments, and has been the principal market for the jewellers of Europe for more than two hundred years. The dark green variety under the name of Brazilian emeralds was imported as early as the middle of the seventeenth century. The production in Ceylon and India has for a long time been large, though in some sections the government restrictions upon the exportation of gems have limited their circulation to the poorest specimens; none of the finest rubellites of Burmah ever reach the markets of the world unless they are smuggled out of the country. A magnificent group, however, was presented to Colonel Symes, in 1799, by the King of Burmah, which was valued at five thousand dollars, and is now in the British Museum. This institution contains a superb collection of tourmalines, from nearly every region where they are found:—pink and crimson, from Ava, comprising a large number of acicular crystals standing like basaltic columns; the colorless, from Elba and the Dolomite Mountains; rose-colored, from Moravia; deep red, purple, and blue, from Sweden; clear light green, from St. Gothard; different shades of brown, from the Tyrol, affording the exceptional instance of having both ends of the prism faceted alike; brown and blue, from the United States; and greenish yellow, from Canada.

Siberia furnishes some magnificent specimens of various colors, including ruby-red, purple, green, and other tints, differing in the arrangement of colors from those of any other locality. They are found in many places in this country, but not always possessing the requisite essentials for jewelry. Dana mentions several sections where they are developed:—Chesterfield, Massachusetts, where they occur of different colors, in granite, but generally opaque, sometimes translucent; in Connecticut, New Hampshire, and Vermont, of a dark color; and in New York, New Jersey, and Pennsylvania, presenting shades of yellow, blue, and green.

The most remarkable deposit of tourmalines in America, and perhaps in the world, was found in Paris, Maine, a few years ago. Mr. Hamlin, in his interesting account of the discovery, says the mine, covering only a few square rods, on the brow of a hill called Mount Mica, is one of the most wonderful found in any country. It has yielded, from an area of thirty square feet, nearly forty varieties, some of them being rare and very beautiful. This depository, which seemed almost like an Aladdin's cave, was accidentally discovered by Hamlin and Holmes, two students, who had been searching for minerals. The broken fragments of crystals scattered about, led to an examination of the premises, which resulted in finding a granite ledge perforated with cavities filled with tourmalines and other minerals. Some of the tourmaline crystals were two and one-half inches in length and nearly two in diameter, and of great beauty, tinted with different colors, chiefly red and green, though some were pink, others white, blue, or yellow, and some transparent specimens were similar to the Brazilian emerald. No other single deposit has yielded so great a diversity of colors as that of Mount Mica; the crystals are said to rival the South American gems in beauty, limpidity, and brilliancy, and are nearly equal to the rubellite of Siberia. They represent the dark green of Brazil, the light green of St. Gothard, the pink of Elba, the light yellow of Ceylon, the blue of Sweden, the white of Switzerland. The Hamlin collection at the New Orleans Exposition included specimens of four different greens, besides red, pink, yellow, and white.

The fame of this natural treasury of minerals at Paris soon became widespread, and attracted thither collectors from different countries, who carried off large quantities of the best specimens to enrich the cabinets of the world. Both the Russian and the Austrian consuls visited the spot and obtained a supply for the museums of St. Petersburg and Vienna. It was supposed in 1865, writes Mr. Hamlin, that the mine, which had been worked to the depth of six feet, had been exhausted, but recent excavations have revealed new cavities holding broken crystals. One of these contained a very remarkable group, which suggested the possibility that there may be still unexplored mines to reward the labors of any one who will diligently seek for them. The mining has been resumed, and the yield in 1882 amounted to more than two thousand dollars. The entire quantity of stones suitable for gems obtained from this region would, it is conjectured, amount to a sum between fifty and seventy thousand dollars.

There has been found a large number of tourmalines at Mt. Apatite, in the vicinity of Lewiston and Auburn, in the same state, differing, however, in general appearance from those discovered in other places, and of lighter colors. Nearly fifteen hundred specimens have been taken from this deposit, embracing colorless varieties, and the light shades of pink, blue, brown, and green; the crystals are generally three, six, and nine sided prisms.

The commercial value of this species of precious stone, except for optical purposes, is small compared with many others, but when it is employed in jewelry, the dark-green specimens are the most desirable.

The Turquoise. — The name of this mineral, it is said, was given to the species on account of its having been introduced into Europe from India through Turkey, a statement suggesting doubt, since Persia is its chief native place, where it is found in the veins of clay-slate crossing the mountains in all directions. The best quality comes from Khorassan, though varieties of less value are found in Thibet, China, Silesia, Saxony, the Isthmus of Suez, Mexico, and the United States. It has been claimed that Russia yields the turquoise, an error easily made, since much of it is cut and polished at Moscow and sold at the great fair of Nijni-Novgorod by Persian and Turkish merchants.

The variety recently discovered at Mount Sinai, Arabia, is unrivalled in its fine blue color when first mined, but it changes its hue in some mysterious manner, a tendency shared by the Persian and other kinds, though in less degree. Antique engraved turquoises are known to have retained their original freshness of color.

The Mexicans were familiar with the use of this substance, and probably obtained it from their own country. It has been found in New Mexico, at Turquoise Mount in Arizona, and in Nevada, in the United States. The specimens from the latter place are of a rich blue, approaching in quality the finest Persian, but are of minute size; those from New Mexico are green, and were highly valued for ornaments by the original inhabitants. The mines in this territory were worked by the Spaniards two hundred years ago, and from them were taken many of these gems found in the crown jewels of Spain.

Some mineralogists have divided the turquoise into what they call "Old Rock," comprising the oriental or superior kind, which retains its color, and is found in irregular masses, and the "New Rock," of a pale tint, inclining to white, a much less valuable and beautiful variety.

The different analyses of the turquoise have hitherto given different results, but phosphate of alumina, oxide of copper, and iron, are invariably present; water, manganese, and lime, have been added to these constituents, though the composition varies in different localities. It is compact, without crystallization and cleavage, infusible, and unaffected by acids, and stands one below quartz in hardness. The colors are a beautiful sky-blue, green, or gray with a slight infusion of green, and are supposed to be due to protoxide of copper; it retains its natural tints by artificial light, and needs no foil to enhance its beauty.

Odontolite, or bone turquoise, sometimes sold for the true article, is composed of the teeth of fossil mammals colored a fine blue by contact with phosphate of iron and copper, differing, however, from the Persian turquoise in the *inky* shades of its colors.

The market value of the turquoise varies considerably; the Persian stones, which are always preferred for jewelry, when of large size and fine tints, bring extravagant prices, though the best specimens are difficult to obtain since they are appropriated by the Shah, and are freely used to ornament the hilts of swords and the handles of knives and daggers. The Persian ambassador to the court of Louis XIV. presented his majesty with a large number of fine specimens, which accounts for their presence in the crown jewels of France.

The turquoise is more frequently used for amulets than any other precious stone, on account of its supposed mystical powers; as a personal ornament it is well adapted for an evening gem, and can be worn with diamonds and pearls.

It was not known to the ancients by its modern name; at least, it does not occur in their writings, though several antique specimens are in the Vatican Museum, and it is frequently discovered among the ruins of Egypt. It was first mentioned by an Arab of the twelfth century. Mr. Streeter thinks it doubtful whether the turquoise was known to the nations of antiquity, but it is pretty certain it was used by the Greeks and Romans, and, probably, by the ancient Persians. It was highly esteemed in the Middle Ages for its remarkable properties, and was believed to grow pale when worn by a sickly person, and to change its colors with the hours of the day, a superstition alluded to by Ben Jonson in his "Sejanus": —

> "Observe him as his watch observes his clock
> And, true as turquoise in the dear lord's ring,
> Look well or ill with him."

It was believed to give warning to its owner of an approaching calamity. Dr. Donne says :—

> "As a compassionate turquoise doth tell,
> By looking pale, the wearer is not well."

Shakspeare represents Shylock as saying he would not have lost his turquoise for "a whole wilderness of monkeys."

Both the *callais* and the *callaina* of Pliny have been thought to be identical with the modern turquoise, since they correspond in some characteristics to it; the callaina, of a pale green color, found among the rocks of Mount Caucasus and in Carmenia (Persia), yielded the best quality. This naturalist relates the curious story that the Persians were accustomed to obtain the turquoise from the inaccessible heights of their precipitous mountains by shooting arrows to detach it from the projecting cliffs and bring it within reach. Theophrastus mentions a fossil ivory with variegated colors of white and blue, probably odontolite, which was used very extensively by the jewellers of his time, as it is by those of the present day. Modern turquoise is sometimes called callaite, or callainite, from the *callais* of Pliny, but this mineral species, though it resembles turquoise in colors, yet differs from it in composition and some of its properties, being inferior in hardness and specific gravity.

Inscriptions, consisting of texts from the Koran, both in Persian and Arabic, were cut on turquoise, but engravings on this stone are not abundant. The most notable are a figure of Diana and another of the Empress Faustina, in the collection of the Duke of Orleans; one in Moscow, which formerly belonged to Nadir Shah, inscribed with a sentence from the Koran; and one in the Florence collection. The best specimens, in the opinion of Mr. King, are the head of Augustus, in the Pulsky collection, and a Gorgon's head, in the Fould. He says many of the ancient intagli and camei cut on this gem are of doubtful antiquity, on account of the perishable nature of the material. A magnificent necklace composed of twelve turquoises of a pale blue, engraved with the twelve Cæsars in relief, was sold at the beginning of the present century, for only one thousand eight hundred dollars.

Lapis-lazuli. — This mineral does not properly belong to the family of precious stones, although it has been used from very early times in jewelry and other decorations, and is supposed to be the sapphire of antiquity, which has been described by the writers of those times as spotted with glittering particles or "shining with golden specks, like a serene sky adorned with stars." Its name, *lapis azul*, from the Arabic, signifies "blue stone."

Haüy considers this mineral identical with the lazulite of mineralogists, but Dana classes them as distinct species, differing in composition and specific gravity, though the precise constituents of the blue variety have not been ascertained beyond a doubt. The lapis-lazuli is regarded by Church, not as a definite mineral, but a mixture of a colorless and a blue substance called haüyne, spangled with minute yellow particles of iron pyrites. It is translucent to opaque, with a hardness varying from five to six, and colors comprising azure-blue, violet-blue, red and green with white or yellow spots, and sometimes white; it loses its beautiful azure by exposure to heat and moisture. Its incapacity to harmonize with other colors constitutes a serious defect in this gem, but for other ornamental purposes it may be used with an agreeable effect. It is found in masses, rarely in crystals, in Siberia, Tartary, China, Persia, Thibet, and other oriental countries, in some parts of Europe, and in Brazil. Marco Polo, during his travels, discovered it on the Oxus, and, on the authority of Haüy, the best quality is obtained from Bokhara and from China, where it is made into various articles for personal ornaments.

This stone has been employed, both in ancient and modern times, for architectural decoration, with striking effect. The Chaldæans and Assyrians employed it with ivory, for the wall surfaces of their magnificent palaces; it was used in the Middle Ages for the embellishment of ecclesiastical buildings, as may be seen in many of the churches of that period, one of the most conspicuous examples being the chapel of San Martino, Naples. An apartment in one of the royal palaces of St. Petersburg has the walls entirely covered with lapis-lazuli and amber. It is used also for mosaic, inlaid work, vases, and for decorating furniture, and was formerly employed for the pigment known as ultramarine, which is at the present time derived from cobalt, or manufactured by artificial methods.

The lapis-lazuli is objectionable for engraving, on account of the pyrites disseminated through the mass rendering it difficult to work. It was, however, used for intagli and camei during the Roman period, though rarely any antique engravings of superior workmanship appear on this material, unless the fine specimen of Alexander the Great, with Apollo, Venus, and Cupid on the reverse, belongs to the age of this conqueror, as is pronounced by some antiquaries, in opposition to King, who thinks it is a production of the Middle Ages.

It is said the antique fashion of cutting fine Persian lapis-lazuli for brooches and pendants has recently been revived.

The Zircon.—This stone is supposed to be the same as the *lyncurium* of antiquity, though Pliny discards the idea that such a gem as the lyncurium was ever known. Ancient intagli are found upon the zircon, therefore it must have been familiar to early engravers by some other epithet, not now identified. The name is thought to come from the Arabic word *zerk*, meaning gem.

It is a rare and beautiful mineral, affording a range of rich and delicate shades, which, for their remarkable play of colors and brilliant lustre, place it next to the diamond as an ornamental stone. The transparent colored specimens and the colorless varieties are used in jewelry. In its physical qualities, it affords an instance of decided double refraction, with a hardness a little above quartz, and crystallizes in the form of double pyramids. The colors are variable — red, green, blue, yellow, brown, gray, amber, all presenting many gradations, and a colorless variety, which, on account of its high refracting power, transparency, and lustre, is often passed for the diamond. Some of the red varieties are remarkable for the vividness of their tints, and have been likened to a flame of fire; the blue and yellow hues are rare. Silica and zirconia, with iron for coloring, form the constituents of this species of precious stone. The zircon, in some of its varieties, has sometimes been taken for essonite; but its composition is quite different from that of the garnet.

It is found in many different countries, including Ethiopia, India, Arabia, Ceylon, Norway, Bohemia, Saxony, France, New South Wales, and the United States though it has not yet been discovered in this country of sufficient size to be of great value as a gem-stone. The best specimens are obtained from Ceylon, New South Wales, and France.

Sometimes the terms zircon, jargoon, hyacinth or jacinth, are indiscriminately applied to the species; whereas the last named stones are only varieties of the zircon. The names hyacinth and jacinth are really the same, the former being Greek, the latter Arabic, and though sometimes applied to varieties of other species, as corundum, vesuvianite, topaz, garnet, and some others, the true hyacinth is the transparent, bright-colored zircon, while the name jargoon is given to the colorless or smoky varieties. A bluish violet gem known to the ancients as hyacinth is supposed to have been the modern sapphire.

The red zircon or hyacinth, remarkable for lustre, resembles the ruby, and the pale yellow, which is extremely brilliant, might be mistaken for the yellow diamond or the topaz; it is even considered superior to the latter for ring-stones, though it is not a favorite in the circles of fashion. It is porous, as may be seen by holding it up against a strong light. Hyacinth is found in rolled pebbles in Ceylon and France.

The jargoon, written also jargon, is a grayish or smoky white zircon resembling the diamond, and considered, until the present century, an inferior diamond. The variety obtained from Matura, Ceylon, where it is called "Matura diamond," is often sold in the bazaars of India for the genuine diamond; it seldom occurs in crystals of more than ten or twelve carats weight.

The Chrysolite. — The "golden stone," as the name signifies, supposed to be the topaz of the ancients, is a title applied to varieties of several different species of gem-minerals, but it is not identical with the chrysoberyl, as has sometimes been represented, differing from it in composition, hardness, and other characteristics. Silica, magnesia, and iron form the largest constituents, which shows that the chrysolite has very little in common with the chrysoberyl. It is one of the softer gems, ranking below seven, is electric by friction, doubly refractive, transparent to translucent, and in color represents various shades of green, yellow, brown, and gray, to nearly white. The kind used in jewelry is a pale, yellowish green stone, cut in the form called step, or as a rose diamond. The chrysolite is said to be the only precious stone set transparent by the ancient Romans, all others being foiled with gold or copper, with the view of enhancing their brilliancy. Sometimes this mineral is found in masses of the size of a turkey's egg, but more frequently in comparatively small crystals.

There is some difference of practice in the classification of the chrysolite, which makes it very difficult to understand where it should belong, if indeed it has any legitimate place at all. It is supposed to be the chrysoberyl of Werner, the cymophane of Haüy, and to claim relationship with the beryl and other gems.

The varieties include the *peridot ordinaire,* in distinction from the oriental, and *olivine,* both depending upon color, which is deep olive-green in the latter, and yellowish green in the peridot. It is known, however, that some modern writers on precious stones consider the names chrysolite, peridot, and olivine interchangeable terms for the same species, while others represent olivine the species, and chrysolite and peridot the names of varieties.

The chrysolite is a volcanic mineral found in Egypt, Turkey, South Africa, Australia, France, Mexico, and the United States; much of that used in Europe is brought from the Levant, but the finest specimens are of Egyptian origin. It is said to occur of very good quality in small pebbles imbedded in the sands of Arizona, Colorado, Montana, and

New Mexico, associated with garnet and sapphire. The variety called olivine is found in meteoric as well as volcanic rocks.

The peridot, a name derived from the Arabic word signifying precious stone, was used in ancient jewelry, and, at some periods later, was considered of more value than the diamond, when it received the title of "evening emerald." It takes a fine polish, attained, however, with extreme difficulty, and by a process known to but few lapidaries, but on account of its softness it is easily scratched.

This precious stone was seldom used by the ancients for engraving, though frequently employed by modern artists for this purpose. There are two Roman intagli executed upon peridot, one of them engraved with the head of Minerva, and the other with that of Medusa.

The Chrysoberyl. — This precious stone, sometimes confounded with the true chrysolite, a mineral of a different chemical composition and inferior hardness, is called by some writers oriental chrysolite. The name signifies "golden beryl"; its varieties comprise the *cymophane*, *Alexandrite*, and *cat's-eye*, terms frequently applied to the species, which adds to the difficulty of a distinct classification of gem-stones. Haüy regards cymophane, chrysoberyl, and oriental chrysolite terms for the same mineral; and King says the chrysoberyl, or cymophane, when cut *en cabochon*, exhibits a peculiar opalescence, when it is called cat's-eye. It was known to the ancients by another name, and the chrysolite was mistaken for the chrysoberyl by the early lapidaries, as it has sometimes been by modern artists and collectors.

This species of gem falls but little below the sapphire in hardness, while in brilliancy it is nearly equal to the diamond, qualities which render it desirable for jewelry, though it has sometimes been relegated to obscurity by the caprices of fashion. Its chemical constituents are alumina and glucina, with traces of some other substances for coloring agents; it is transparent to translucent, occasionally opalescent internally, displays a remarkable play of colors, and is double refracting. Its glittering crystals exhibit different shades of green, brown, yellow, and white colors; the transparent yellowish specimens, cut with facets, and the opalescent varieties, *en cabochon*, are those most frequently used for ornamental stones.

The term *cymophane*, meaning "to appear like a wave of light," is given to a variety of the chrysoberyl when it has the appearance of enclosing rays of light, which seem to be floating in the interior of the stone, a phenomenon supposed to be the result of blue reflections emanating from a milky-white substance; or the gem may be compared to a drop of water with a beam of light imprisoned within. Examples of this variety of the chrysoberyl are seen in the South Kensington Museum.

Alexandrite.—This red and green variety of the chrysoberyl was named for Alexander I., Emperor of Russia, who adopted these hues for the imperial colors. It affords a good example of dichroism, presenting a dark green by daylight, which changes to a columbine red in the evening or by artificial light. Its discovery in the Urals is of recent date, though it is said about one-third of all those sold are from Ceylon, a large per cent of them weighing over sixty carats.

Cat's-eye.—This variety of the chrysoberyl is the true cat's-eye and entirely unlike the chatoyant quartz erroneously called by this name. It seems to be a sub-translucent form of cymophane, and partly the result of art or the form of cutting; its peculiar play of colors is attributed to minute internal striations. It was called by some of the nations of antiquity "oculus solis," eye of the sun, and is at the present day a favorite in China, where it commands a high price. The cabinet of Vienna contains a cat's-eye of a yellowish brown color, which measures five inches in length. The sum of five thousand dollars, says Emanuel, was paid recently by an English nobleman for a cat's-eye of extraordinary size and beauty.

Iolite. — This mineral is known by several names: iolite ("violet stone"), dichroite ("two colors"), and sapphire d'eau ("water sapphire"). It is a beautiful and interesting stone, remarkable for its play of colors, and is occasionally used for gems. When cut, the crystals present different shades of blue, red, brown, and yellowish gray, according to the direction in which they are viewed. Its chemical constituents are silica, alumina, magnesia, and iron; its hardness, exceeding that of quartz, and its fine color, render it a desirable ornament. The transparent specimens, called sapphire d'eau, resemble the true sapphire, and are sometimes sold for this precious stone; those from Ceylon, of intense blue and used in jewelry, are very complex in their composition and conspicuous for their dichroism. Iolite is found in Greenland, Scandinavia, Bavaria, Spain, Tuscany, Ceylon, and in some localities in the United States.

Kyanite. — This name signifies dark blue color, though the mineral affords examples of other colors,— white, gray, green, and black. It is sometimes called disthene, meaning "twice," or of two kinds, because it possesses both positive and negative electricity. The sky-blue variety, when transparent, is occasionally used in jewelry, and is sometimes sold by oriental lapidaries for sapphire. Its composition is similar to that of iolite, with the exception of magnesia, but its hardness is inferior; it is found in many different places in both hemispheres.

Apophyllite. — This substance is sometimes classed with gem-minerals, but its deficiency in hardness, ranking only from 4 to 5, renders its use for jewelry very doubtful. The name is derived from its tendency to exfoliate under the blow-pipe. It is generally transparent, white or gray, occasionally tinged with green, yellow, or red, and has a pearly lustre resembling the eye of a fish, whence it is sometimes called "fish-eye stone." It occurs in India, the Harz Mountains, Greenland, Iceland, Mexico, the United States, and other localities.

CHAPTER XVII.

LABRADORITE AND OTHER GEMS.

THE opinion has been advanced that labradorite was unknown until about a century ago, when it was discovered at Labrador, from which it received its name, but this statement can apply only to the inhabitants of the Eastern Continent, since Indian ornaments made of this stone were found by the Spaniards during their invasion of America. It occurs in volcanic matter, and forms a constituent of porphyry, verde antique, and some other rocks. This mineral is complex in its chemical nature, comprising a variety of substances — silica, alumina, lime, soda, iron, and sometimes magnesia, potash, and water, and, notwithstanding its comparative softness, admits a fine polish, which constitutes it a beautiful and desirable decorative stone. It is employed to some extent for jewelry.

Labradorite is of a grayish color, but, when seen in certain directions, it displays chatoyant reflections, usually of blue, green, violet, and gold, "equal to those which ornament the most beautiful butterfly." There is a difference in this respect among the numerous specimens of this mineral to be met with ; in some instances, red, gray, amber, and orange tints are conspicuous, while in others the play of colors is wanting. This iridescent quality is due to its peculiar structure, which, like the opal, encloses fissures, occupied, it is conjectured, from experiments with the magnet, by thin films of iron, and perhaps minute scales of other minerals, thus increasing the peculiar play of brilliant colors seen in this beautiful substance. A variety of labradorite has been incorrectly called saussurite, and jade or nephrite. As an instance of one of the curious freaks of nature, it is mentioned that a slab of labradorite found in Russia bore an image of Louis XIV., of France, wearing a crown of pomegranate, with a border displaying all the prismatic colors, and a plume of bluish tint. This marvel of natural painting was owned by a Russian noble, who refused to part with it for two hundred and fifty thousand francs.

Jade. — This is a generic term, including various mineral substances, as nephrite, saussurite, and others. The names jade and nephrite have the same origin, from words signifying "kidney," given to these minerals from the opinion that they were efficacious in diseases of the kidneys. Some mineralogists consider jade distinct from jadeite, while others class it as a translucent variety of zoisite, the *tenos* of Pliny. King thinks it doubtful whether jade was known to ancient classic writers at all, since it was imported, with its oriental name, into Europe from the East, by the Portuguese.

Its composition includes a variety of substances, silica and alumina being the principal. The color passes from dark green to cream-white. The Chinese variety is generally of a light green or bluish white, while that from New Zealand is a rich, dark green.

This mineral is also found in Egypt, Australia, Switzerland, and, it is thought, in Alaska, east of Point Barrow, since implements made of dark green jade were found in this vicinity by the members of the Alaska Expedition. Ornaments made of this material, obtained from burial mounds in Nicaragua and Costa Rica, agreeing in color, hardness, and specific gravity with Asiatic jade, have suggested the idea that the natives of these states obtained their supply from the East.

The Chinese employ jade for vases and other articles of a similar character, of various shapes and sizes, which are to be seen in great numbers at all international expositions, and in the shops of dealers in such wares. Translucent specimens of the best quality are used for ear-pendants, which Indian lapidaries cut with great skill, showing remarkable lightness and delicacy of workmanship. It is probable the stone hardened after it was taken from the mines and cut, since antique objects made of jade are so hard that no material but the diamond can scratch them. For want of brilliancy, this substance does not rank high as an article of jewelry, but it is used with beautiful effect for ornamental vases. The Caribees, as related by Humboldt, wore jade amulets cut in the shape of Babylonian cylinders, — an interesting fact to the antiquary, which might, possibly, throw some light upon the origin of the native races of the New World. The jade found in Mexico, highly prized by the Aztecs, was considered by the Spaniards a kind of emerald.

Nephrite. — This mineral, identical with the *pietra de hijada* of Mexico and Peru, is a tough, compact tremolite, introduced into Europe after the conquest of these countries, and is similar to that found in China and New Zealand. The bowenite of Rhode Island resembles it in appearance, though it differs in composition. Nephrite is rich in color and very tough and hard. What is called soft jade is a kind of steatite, or soap-stone.

Amazon-stone. — Microcline, or Amazon-stone, was the axe-stone jade from South America, which Dana classes as a variety of orthoclase; it is pale green, with nacreous reflections, and susceptible of a high polish, but is very brittle. The name was given to it from the supposition that it was first discovered on the Amazon river; but Haüy says it is a misnomer, as it is found only in the Russian Empire and in Greenland; but since his day it has been known to occur in other places. It was discovered at Pike's Peak, Colorado, in 1875, in a locality retired from any travelled road, imbedded in a kind of graphic granite, in pockets, at a depth of eight or ten feet, and extending over a very limited area. Amazon-stone has also been found in Maine, Pennsylvania, Virginia, and North Carolina. It occurs in large crystals of a light bluish green color passing to a dark emerald green, while some specimens are yellowish, flesh color, or white.

Saussurite. — This mineral, discovered in the vicinity of Lake Geneva and named for De Saussure, appears to be a variety both of zoisite and labradorite; it is of a pale green color, passing to nearly black, and is employed as an ornamental stone.

Malachite, Azurite. — Malachite is classed by Haüy with the inferior precious stones, and, though hardly worthy the name of gem, it has been used for jewelry, both in ancient and modern times.

Theophrastus alludes to a species of stone found in copper-mines, and called false emeralds, which may have been either malachite or chrysocolla, since both contain a large per cent of copper and are found in or near such mines. The *molochitis* of Pliny, obtained from Arabia, of a deep green and nearly opaque, was, without much doubt, the same as the malachite of the present day. The name of this species of mineral is derived from *moloche*, or *malache*, signifying "mallow," conferred in reference to its color, which resembles the hue of that vegetable.

Malachite is a hydrous carbonate of copper — some scientists say, the product of decomposed minerals containing copper.

The green carbonate of copper receives the name malachite; the blue carbonate, that of azurite. There is only a slight difference between these two species except in color — the constituents being the same, with a little variation in their relative proportion. The azurite presents different shades of azure — blue to a deep blue; that found at Chessy, France, has been denominated Chessy-copper, or chessylite. The bright green malachite is spotted and banded with tints of paler green; it is very soft, having only a hardness of three and one-half, but on account of its remarkable play of colors and capacity for polish, when found in masses of sufficient size, it yields an excellent material for ornamental vases, boxes, tables, and other articles of similar use.

These minerals are the native products of many different countries, but the Russian mines yield the best quality of malachite, especially those of Prince Demidoff, in Siberia, whose fine collection, exhibited at the Exposition of 1851, first directed public attention to this material for decorative purposes. Though it is developed in many of the states and territories of this country, and sometimes in considerable masses, it has not yet been mined for art uses. A rare specimen of malachite and azurite combined, discovered in Arizona, was seen at the New Orleans Exposition.

The mineral collection of St. Petersburg is said to comprise one specimen weighing ninety pounds. A remarkably beautiful vase of malachite, made by order of the Emperor Nicholas for the King of Prussia, is in the Museum of Berlin; and another magnificent vase, presented by the same royal donor to Sir Roderick Murchison, the celebrated Scotch geologist, is contained in the Geological Museum of London. Ancient engravings on malachite are seldom met with, though one specimen of the kind is said to be found among the Pulsky collection of gems.

Moonstone. — It is no easy matter to class this chameleon-like gem, since it claims kindred with so many species, and passes under so great a number of names. It constitutes a variety of orthoclase, albite, and oligoclase, all of the feldspar group, while the ancients applied the name to selenite or gypsum. It has been called cat's-eye, a name given to four other different stones; argentine, a pearly lamellar calcite; œil de poisson (fish's eye); water or Ceylon opal, and adularia. The name hecatolite has been given to it for the same reason that it receives that of moonstone — namely, because it was thought to enclose the image of Luna, one of the forms of the threefold goddess Hecate. The moonstone of Dioscorides, which he calls moon-froth, was probably crystallized gypsum.

This mineral exhibits a silvery or pearly light, not unlike that of the moon, and in some of its varieties it resembles ice; it is opalescent, white, grayish, yellowish, or reddish in color, and as an ornamental stone is very fashionable in some countries, where it is sold in large quantities. Moonstone of good quality, resembling that from St. Gothard, is found in Pennsylvania and Virginia, but the best variety comes from Ceylon, which yields some fine gems, known to measure more than an inch in length.

Sunstone. — The aventurine oligoclase called sunstone exhibits prismatic reflections of a golden or reddish hue, the result of minute disseminated crystals of hematite, göthite, or mica, and is sometimes used for jewelry. The term aventurine applies to any mineral spangled with scales of some bright substance, and not to a species or to any particular variety of a species, — as aventurine quartz, aventurine feldspar; but sometimes the name is used in an indefinite manner, as when sunstone is called oriental aventurine. A variety of this mineral, from St. Gothard, passes under the name of adularia, which differs from the orthoclase adularia. Sunstone of very good quality is known to occur in certain localities in the United States, but those of the finest quality are brought from Archangel, Russia.

Adularia. — This mineral is a transparent variety of orthoclase, which is characterized by pearly or opalescent reflections, and a play of colors resembling labradorite; it was named for one of the peaks of St. Gothard, where it is found, and is identical with the valencianite of Mexico. The opaque, green adularia is called amazon-stone — a term applied to varieties of other species; and it is sometimes used for jewelry, under the name of moonstone.

Phenakite. — This name is equivalent to "deceiver," and was conferred upon this species on account of its having been frequently mistaken for other minerals. The white variety may easily pass for the diamond if the play of prismatic colors is very conspicuous, as not unfrequently happens. It is sometimes transparent, but oftener translucent or clouded. The colors do not embrace a wide range, consisting of brown, and a bright wine-yellow inclining to red: its hardness is superior to that of quartz, and the crystals, often of large size, are double-refracting. Silica and glucina are the only substances which enter into its composition.

The best specimens come from the emerald and chrysoberyl mines of Asiatic Russia, and some stones of suitable quality for jewelry have been discovered at Pike's Peak, Colorado. As a gem, phenakite is rare, but may be found in collections; the British Museum contains some fine crystals.

Zonochlorite. — This mineral species was discovered by Dr. A. E. Foote, in 1868, at Neepigon Bay, on the north shore of Lake Superior, in an amygdaloid trap, associated with some other minerals. Its constituents are lime, silica, alumina, soda, and water; it does not crystallize, but is found in masses with bands arranged in concentric layers of different shades of dark green. With a hardness, in some specimens, equal to quartz, it admits of a good polish, — a quality, together with its fine color, constituting it an agreeable gem-stone. The name is a combination of three words, meaning "banded green stone."

Diopside.— The name diopside, meaning "double appearance," was given to a variety of pyroxene, on account of its dichroism; its range of color includes white, brown, and various shades of green resembling green tourmaline, or green epidote. Though softer than quartz, the transparent crystals are cut for gems. The Tyrol is the best known locality for this mineral on the Eastern Continent, but fine specimens, weighing from six to fifteen carats, comparing favorably with the imported stones, have been discovered at De Kalb, New York, said to be the only place in this country where they have been found.

Dioptase. — The mineral known as dioptase, signifying "to look through," regarded by some lithologists as a green variety of the beryl, is classed by Dana as a species, closely allied, however, to that precious stone, though differing in hardness, specific gravity, and chemical composition, which consists of a large proportion of oxide of copper. The crystals are transparent to translucent, and possess the quality of double refraction in a high degree; it closely resembles the emerald in appearance, for which it is sometimes sold. It is very limited in its natural distribution, occurring almost exclusively in the Kirghiz Steppes, Siberia, where it was discovered and named achirite.

Epidote. — The epidote, occasionally employed as a gem-stone, is sometimes denominated green schörl. The predominant color is a peculiar yellowish green, seldom met with in other minerals, which frequently passes into red, yellow, gray, and black. Its chemical constituents include silica, alumina, iron, and lime, with a small per cent of other substances in some varieties; it has a scientific interest, on account of its very distinct pleochroism and strong double refraction. Although quite abundant in some of its forms in the United States, yet it seldom exists here of a quality suitable for an ornamental stone.

Euclase.—This mineral was first brought to Europe from South America, in 1785, but it has since been obtained from the Ural Mountains. It has been classed with precious stones by certain writers; though, in spite of its transparency, hardness, and capacity for polish, it is rarely used as a gem, on account of its brittleness, a quality which suggested the name—"easy to break," given by Haüy. Silica, alumina, and beryllium form the largest constituents. Pale green, blue, and white varieties are included in this species.

Crocidolite.—The name crocidolite, from *krokis*, "woof," was given, in allusion to a fibrous structure, similar to that of asbestos, to a soft, opaque mineral, of different colors, occasionally employed in jewelry. It presents an array of colors varying from gold to yellowish brown, leek-green, dull red, deep blue, and greenish blue, and is one of the soft stones, not exceeding four in hardness. It is developed in Greenland, Norway, the Vosges Mountains, India, and on the Orange River, South Africa, where all those used for jewelry are found. This gem is sometimes called "tiger's eye," because when cut and polished it exhibits the chatoyant lustre of that animal's eye.

Titanite.—Titanite, called also sphene (which means a "wedge"), has been employed, to some extent, for a gem-stone.

One of its constituents is titania, as is intimated by the name of the species. Its color varieties afford green, yellow, gray, brown, and black crystals, which exhibit a great difference of forms, and strong refracting and dispersive powers; it holds a middle rank for hardness. Fine yellow specimens, yielding beautiful gems, are obtained in Switzerland and other regions, while black and brown varieties are plentiful in the United States, but not of a quality suitable for the uses of jewelry. In some of its varieties it resembles fire-opal.

Lepidolite. — Several countries of Europe, as well as some localities in New England, afford this species of mineral, which is frequently used for making ornamental boxes, vases, and other fancy articles, and sometimes for personal ornaments, though it is not, strictly speaking, a precious stone. The name, meaning "scale stone," is due to its peculiar structure. It is very soft, holding a rank of only two and one-half in the scale, translucent, and is very complex in composition. The range of colors includes white, rose, yellow, and different shades of gray. The Indian lapidaries cut the crystals of lepidolite with facets or *en cabochon*, which they sell for sapphires.

Chlorastrolite. — This name is composed of three Greek words, signifying "green-star-stone." The mineral occurs in small, rounded, water-worn pebbles which have come from trap, on the shores of Isle Royale, Lake Superior, and is exclusively an American gem. It is opaque, of a light bluish green color, presents a stellated or chatoyant appearance, and, with a hardness of five to six, takes a fine polish. It is sold for cabinet use or gem-stones, the annual profits from this source reaching, it is estimated, from two thousand to three thousand dollars. Specimens one inch in length and of fine color have been valued as high as fifty dollars each.

Axinite. — This substance is very rarely used in jewelry,

though it makes an agreeable ornament when cut *en cabochon.* The name signifies an "axe," given in allusion to the form of its crystals. Axinite is transparent to translucent, and affords blue, plum-colored, brown, and light gray varieties, and sometimes crystals are known to combine three colors, brown, blue, and green, when viewed in certain directions. It possesses sufficient hardness to admit of a good polish, but its quality for a gem-stone is impaired by its brittleness. It is said this species and the tourmaline are the only minerals used for precious stones, containing boron, which in axinite is found only in small quantities in a few specimens. Though a native of this country, the best varieties are brought from France.

Vesuvianite, or Idocrase. — The "Gem of Vesuvius," or the "Hyacinth of Vesuvius," as this species is sometimes called, is identical with the modern idocrase, which means "see" and "mixture." Vesuvianite is the more appropriate name, since it was first obtained from the ejected rocks of Vesuvius and Somma, though it has since been found in Piedmont, Norway, the Urals, Hungary, Canada, and the United States. Haüy calls the pale yellow variety volcanic chrysolite, and the yellowish green, volcanic hyacinth. This species is similar to the garnet in composition, but differs from it in the form of its crystals; it is non-electrical, and possesses refractive powers only in a feeble degree. With a hardness of six and one-half, and an array of colors including different shades of green, yellow, orange, blue, brown, and, occasionally, black, it constitutes quite a desirable ornamental stone. When cut, it is sold at Naples and Turin under the name of "Gemme di Vesuvio," chrysolite, or hyacinth, according to the color. A pale blue variety is called cyprine.

Obsidian. — Iceland agate and volcanic glass are both terms which have been applied to obsidian. Church calls it melted orthoclase, and Dana says it often consists of a mixture of labradorite, augite, chrysolite, and iron. It is not found in crystals, but occurs in globules or masses, affording reddish brown, green, yellow, and black varieties. The last resembles black spinels, tourmalines, and garnets, only they are more translucent. On the authority of Pliny, it was named for Obsidius, who claimed to be its discoverer, and was used by the Romans not only for jewelry, but also for mirrors which reflected the shadows of objects. The Emperor Augustus dedicated four elephants cut in obsidian to the gods in the Temple of Concord, at Rome. An obsidian statue of Memnon was found in Egypt during the reign of Tiberius, thus proving its great antiquity for art purposes. It was used by the ancient Peruvians and Mexicans for personal ornaments, mirrors, and cutting instruments, and when faceted it is still employed occasionally in jewelry. Moldarite, or bottle-stone, of Moravia, is green obsidian. The American variety, found in large masses in California and other Pacific regions, is seldom used for ornamental purposes.

Lodestone. — This substance, called also magnetite, has been employed in art for intagli, especially those representing Gnostic subjects. Its wonderful magnetic powers induced Dinocrates, a celebrated architect in the time of Alexander the Great, to begin a temple dedicated to Arsinoë, wife of Ptolemy Philadelphus, the roof of which was to be made of lodestone, so that the iron statue of the queen might remain suspended, as if floating in air, — a plan the artist did not live to complete. This story may have suggested the fiction about the coffin of Mahomet.

Claudius mentions two statues made of lodestone, one of Venus and the other of Mars, which, when placed in the same temple together, were attracted towards each other by a mutual magnetic influence. Pliny, in his description of this stone, says, "Nature has bestowed upon it both feet and intelligence." It was known by the names of *magnes*, for its discoverer; *sideritis*, meaning "iron earth"; and *heraclion*, either from Heraclea, a place, or from Hercules, on account of its great power. Of all the numerous varieties, the Ethiopian was considered the best, and was equal to its weight in silver. The species was believed to be efficacious in ophthalmic diseases.

Steatite.—White steatite, a variety of talc as well as of saponite, or soapstone, is carved into beautiful ornaments at Agra, India, and has sometimes been used for personal decoration, notwithstanding its deficiency in nearly all the requisites of a gem-stone.

Selenite, a kind of gypsum, regarded as suitable for necklaces, bracelets, and other articles of the toilet, received the name from the goddess Selene, and is supposed to have been the moonstone of Dioscorides. A rare and beautiful variety, of a reddish yellow color is found in Russia.

Vulpinite, a variety of anhydrite, discovered at Vulpius, Lombardy, is cut and polished for jewelry, and may be classed with the inferior gem-stones.

Pyrites, sometimes inaccurately called marcasite, was extensively used in Europe for personal ornaments during the eighteenth century, and was employed by the early Mexicans with turquoise and obsidian for mosaic work. A mask with eyes made of pyrite, brought from their country, is seen in the British Museum. At one time this mineral was used in France for gem-stones set after antique models, but they have little value at the present day.

Spodumene was not recognized as a gem until recently, when brilliant, transparent crystals, imported into Europe from Brazil, excited the attention of lapidaries, who cut and polish them for ornamental stones. The name is derived from *spodes*, "ashes," on account of its assuming an ash color under the blow-pipe. Silica, alumina, lithia, and iron, with traces of some other minerals, are found in some of its varieties. Its hardness, equal to that of quartz, and its delicate green color, sometimes shading upon red, constitute it a desirable material for the jeweller's art. It is found on the Eastern Continent, and in various places in the United States. Green, transparent crystals have been discovered in North Carolina, and amethystine colored specimens in Connecticut.

Hidennite, or lithia-emerald, a new variety of spodumene, was discovered on the farm of Mr. Warren, in Alexander County, North Carolina, in loose crystals, sparsely scattered over the surface of the soil. These were shown to Mr. W. E. Hidden, of New York, a collector of minerals, who leased the grounds and carried on a systematic exploration in 1880, which resulted in his discovery of the mineral *in situ*. It occurs, says Dr. J. L. Smith, in metamorphic rocks, generally gneiss or mica-schist, in veins of hard kaolin, which have been examined to the depth of more than twenty feet without revealing any change of character. The associated minerals consist of quartz, mica, rutile, orthoclase, and beryl. When first discovered, the crystals were supposed to be diopside; but when subjected to a blow-pipe test, they proved to be a new variety of spodumene, to which the name of hiddenite was given by Dr. Smith, in honor of Mr. Hidden, whose successful operations developed this rich mine.

When cut and polished, this beautiful gem resembles the emerald in brilliancy and lustre, and has even been thought to surpass it in these qualities; but in vividness of color it falls below. It is always transparent, and ranges from colorless, a very rare variety, to a deep green, which is generally more vivid at one end of the crystal than at the other. It is said to be worth from thirty-two dollars to two hundred per carat, and the demand at that price exceeds the production.

Chondrodite, a silicate of magnesia and iron, is known to occur in granular limestone in Finland, Sweden, the Urals, Scotland, Canada, and the United States. The name, meaning "a grain," is due to its granular structure. Its hardness, translucency, and range of colors, embracing white, yellow, red, green, brown, black, and gray, have placed it in the rank of gem-minerals.

Bowenite and *Williamsite*, both varieties of serpentine, have been employed for ornamental uses. The former, found in Rhode Island, is a rich green stone resembling nephrite, and was once known by this name. Williamsite, of an apple-green color, is sometimes used as a substitute for jade.

Thomsonite is developed in lava and some metamorphic rocks in various countries; that found near Lake Superior exhibits a flesh-red, banded with green, red, and white, affording an attractive mineral for ornament.

Willemite, said to receive its name from the King of the Netherlands, is a silicate of zinc, and is found in Greenland, the United States, and some European countries. Transparent specimens display rich brown, red, yellow, and green colors; a variety discovered in New Jersey has yielded gems of several carats weight.

Rutile, sometimes called red schörl, is of a reddish brown passing into red, and occasionally into yellow, violet, blue, black, rarely green. It has been found in North Carolina and Georgia of a compactness and lustre suitable for gems. The dark-colored specimens of the former State are thought to resemble more closely the black diamond, in hue and lustre, than any other gem, while those of the latter, after cutting, more nearly approach the garnet in the tone of color.

Sagenite, rutile in quartz, called also Love's arrows, and Venus'-hair stones, received the name from a word meaning "a net," on account of the acicular crystals it encloses. The colors are rich red, yellow, and different shades of brown, which afford a pleasing effect either by natural or artificial light.

Monazite is a very rare mineral, as the name, "to be solitary," indicates, though it has been found in several localities — Connecticut, New York, and North Carolina — in this country, and in the Urals, Norway, and Colombia, South America. Its composition is very complex, including lime and several acids and oxides, while the colors are less diversified, affording hyacinth, yellowish, and clove-brown specimens. It is not usually classed with precious stones, though it is sometimes used for gems.

Euchroite, meaning "beautiful color," has been considered a gem-mineral by some writers, probably on account of its transparency and bright green hue, but its softness renders it objectionable for jewelry. It resembles dioptase, and is found in Hungary.

Barite, heavy spar, developed in several localities in the United States and in Europe, has been sometimes called "Bologna stone." It is white, frequently passing into other colors, with high specific gravity, but low in the scale of hardness; it sometimes occurs in transparent crystals of gigantic size, and when cut and polished it affords a beautiful stone for ornamental household articles.

Hematite. — The name signifies "blood," and the red hematite is supposed to be the bloodstone of Theophrastus, who says it seems to be concreted blood. Though it can hardly be classed with precious stones, yet it was extensively used for engraving by the ancients, the Island of Elba affording beautiful crystals of this mineral in Ovid's time, as it does at the present day. It is so generally diffused that it is needless to specify any localities in which it is found.

Diaspore. — The name of this mineral signifies "to scatter," given on account of its tendency to decrepitate before the blowpipe. It is translucent, with a hardness of six to seven, and embraces quite a variety of colors — greenish, yellowish, brown, gray, and white; some crystals are violet blue when examined in a certain direction, reddish blue in another, and green in a third. Its constituents are alumina and water, with a small per cent of iron and silica; it bears a resemblance to topaz, and might be used as a gem-stone. It is found in several foreign localities, and in Massachusetts and Pennsylvania.

Andalusite, chiastolite, or macle, is sometimes called "cross-stone," from the figure of a perfect cross which some specimens afford when cut transversely, of a different color from the rest of the stone, which is usually rose, violet, gray, whitish, brownish, or greenish. It affords an almost infinite variety of markings, which render this mineral a beautiful ornamental stone. It was called andalusite from Andalusia, in Spain, where it was first discovered, chiastolite from the Greek letter χ (*chi*), and macle from a Latin word signifying a spot. It is found in several foreign localities, and, as a native mineral, in Maine, Massachusetts, Pennsylvania, and California.

Octahedrite. — Crystals of this species, from Brazil, of a beautiful blue color, are of such remarkable brilliancy as to be often mistaken for diamonds; those of inferior quality have been found in North Carolina and Rhode Island. It exhibits shades of brown, passing into deep blue or black, and is of sufficient hardness to admit a good polish.

Agalmatolite, a name derived from "image" and "pagoda," is a variety of pinite, with colors inclining to gray, brown, green, and yellow. It is used by the Chinese for carving images, pagodas, and similar objects.

Several other minerals, in some of their varieties, have a doubtful rank among precious stones, as *orangite*, so called from its color, obtained at Nancy, France; *rhodochrosite*, meaning "rose color," found in Hungary and some other European countries, and in certain localities in the United States; and *crocoisite*, written also crocoite, signifying "saffron," affording specimens with different shades of hyacinth red, discovered in Hungary and Siberia.

Murrhine. — A furious war, says King, has been waged by archæologists about the real nature of this substance, which has been identified, with a good deal of assurance, with various substances, as onyx, agate, opal, fluor-spar, and porcelain. Judging from the description of Pliny, there are evidences that it was a natural, rather than an artificial, production; though Propertius mentions "numerous vessels, baked on Parthian hearths," which are supposed to refer to those made of "murrhina." Whatever may have been the nature of the material, it was, according to the Roman naturalist, greatly diversified in hues, and wreathed with veins of purple and white, passing into flame-color. It was first brought to Rome by Pompey, after his eastern conquest, when numerous murrhine vases and cups, the spoils of his victories, carried in his triumphal procession, were dedicated to Jupiter Capitolinus.

Articles made of this substance brought fabulous prices; a single cup was sold for seventy thousand sesterces, and a basin for three hundred thousand; while Nero paid one million sesterces for a cup — a deed well worthy an emperor, the father of his country, satirically remarks this writer. One individual, of consular rank, owned an innumerable collection of murrhine vessels, which, when seized by this emperor, after the death of the owner, filled a theatre in his palace garden beyond the Tiber. As murrhine must have been something more durable than porcelain, it is reasonable to suppose that some of these vessels, either entire or in fragments, would be found in ancient ruins; but the only articles discovered answering the description are made of agate, which are so abundant as to leave no doubt of their extensive use in ancient Rome; while there are no fragments of fluor-spar ever known to have been seen among the remains. The preponderance of evidence, therefore, seems to be in favor of agate.

Coral. — As a substance employed in art, coral has a very high antiquity, having been a valuable article of merchandise with the ancient Tyrians, who imported it from Syria, as stated by the prophet Ezekiel. It was highly prized among the Greeks, who consecrated it to Jupiter and Apollo as one of their richest offerings. They had a tradition that it was formed from the blood of Medusa, whose head Perseus hung on a tree near the sea, when the coagulated drops, transferred to the water by the nymphs, became coral. The name is from *korallion*, signifying "maiden daughter of the sea."

The true nature of the coral was not understood until discovered by M. Peysonnel during the first part of the eighteenth century. Previous to that event, contradictory and absurd theories prevailed in regard to its source and character.

Corals, like pearls, have an animal origin. It must be admitted that the precious coral, or that kind used in jewelry, has a long lineal descent, having been derived from a single species, if paleontologists can be relied upon, the Corallium rubrum, family Sargonidæ, order Alcyonia, class Actinozoa, sub-kingdom Coelenterata. The black coral belongs to another order, the Zoantharia. Besides lime, its principal constituent, magnesia, iron, and organic matter enter into its composition, but the exact nature of the coloring agent is not assured.

The home of the precious coral is the Mediterranean coast, particularly on the African shore, where it is obtained with great difficulty by means of nets and drags, the beds, in some instances, being seven or eight hundred feet below the surface of the sea. The business is now principally confined to Italian and Maltese traders, although the French, as early as the middle of the fifteenth century, were engaged in coral fisheries. The traffic carried on in this commodity is said to be immense; vast quantities are yearly exported to China, India, and Persia, where it is extensively employed for various objects. One house alone in Naples exported to Calcutta in a single year forty thousand dollars worth of coral, and the total annual amount to India, from this city, was one million dollars. It is used by the orientals both for religious and secular purposes; the Brahmins employ it for rosaries, and the Japanese for personal ornaments.

Coral is a great favorite in Spain, Italy, and the West Indies. When employed for camei, the rough outside of the shell is cut away leaving the smooth inside for the beautiful background on which the figures rest. The delicate pink coral is preferred to the red, and affords a great variety of shades,—one hundred, according to Dieulafait, having been recognized at Marseilles alone. This substance has been considered, and is still believed by the credulous, to be of great importance for amulets, in consequence of its remarkable medicinal properties. It is frequently imitated by bone, horn, and ivory, stained with cinnabar, and an artificial article is manufactured from gypsum and a kind of gum colored by certain pigments. The commercial value of the genuine coral varies, and is dependent, to a great degree, upon the color.

Amber, called in science succinite, was the *succinum* or the *electrum* of classic times. It received the latter appellative on account of its repelling light bodies, whence our word electricity. Various and fanciful have been the theories about the origin of this substance, from early times down to the present, and it is still claimed by some theorists that it is not yet understood. The ancient Greeks, who were always ready to give an explanation of all the phenomena of nature, ascribed its origin to the grief of the sisters of Phaeton, who, after his untimely death, were changed into poplars on the banks of the Evidanus, the modern Po, and expressed their emotions by weeping tears of amber. At the mouth of this river were the Electrides Insulæ, or Amber Islands. Pliny alludes to this legend, but takes the learned Greeks to task for their credulity or imposition about the source of this substance, and proceeds to give what he believes to be the true origin, that is, the gum or juice of a pine-tree,—hence the name succinum, from a word signifying juice. In regard to the "Electrides Insulæ," he says those so ignorant of geography should not write upon the subject.

Amber is supposed to have been unknown to ancient orientalists, since no traces of this substance have been discovered in Egyptian, Chaldæan, or Assyrian remains, though it constantly occurs in the tombs of Latium, Etruria, and Præneste, and was known to the Greeks as an ornament in Homer's time, who refers to the gold and amber necklace presented to Penelope by one of her suitors. At a later period Thales noticed its property of attracting or, rather, repelling light bodies; and Theophrastus speaks of amber in his "History of Stones." Tavernier calls it a certain congelation, made in the sea, which resembles gum, and tells us that the Chinese grandees burned it at their feasts for perfumery, on account of the agreeable odor it exhaled during combustion. It is possible this traveller may have mistaken ambergris, a product, it is supposed, of the sperm-whale, for amber.

It is conceded by modern scientists that it is a fossil resin or gum, derived from an extinct species of pine and other plants, of the Tertiary period. Sir David Brewster *established* the fact of its vegetable origin, which had been conceded eighteen centuries before his time. That it was once a viscous fluid is proved by the insects and plants imprisoned in the substance. There have been found one hundred and sixty-three species of insects entombed in amber, many of them identical with those of the present day; while the plants are different from the vegetation now found in the regions in which it occurs. It is very soft and light, and possesses a remarkable negative-electrical property; it is transparent to translucent, and affords yellow, reddish, and whitish varieties.

The Baltic, the Urals, Switzerland, France, England, Sicily, and some other places of Europe, yield this substance; it also occurs in several localities in the United States, especially in Massachusetts and New Jersey, but in small quantities. A large part of the amber of modern times is obtained from the Prussian Baltic, where the government protects the monopoly of the trade with very stringent laws. The yellow amber of Dantzic, it is estimated, yields from fifty thousand to eighty thousand francs annually; even in Tavernier's day it was farmed out by the Elector of Brandenburg for more than twenty thousand crowns a year. Sometimes it is found in large masses; Pliny mentions a specimen at Rome weighing thirteen pounds, and there is one in Berlin which weighs eighteen pounds. Mr. King speaks of an *elastic* amber ring brought from Egypt; this substance, boiled in turpentine, can be reduced to paste and moulded into any form desired, and it is possible the Egyptian ring may have acquired its elasticity by a similar process.

The Greeks and Romans held amber in high repute for personal ornaments, on account of its beautiful color. Great quantities were introduced into Rome during the reign of the Emperor Nero, which were obtained from the German tribes. This despot, in some verses written by himself, called the hair of his wife amber-colored; directly red hair became then, as it would now under similar circumstances, all the rage among the ladies of the imperial court; and numerous were the devices to secure the fashionable color.

Shakspeare, as well as later writers, alludes to amber ornaments, which appear to have been the general favorite in England at one time. Yellow varieties, cut in facets, have been prized for bracelets and necklaces, and are now employed in some countries for beads. In the East they are often inlaid with gold and precious stones.

Jet. — Though jet, like amber, is of vegetable origin, and not a precious stone, or a stone at all, yet it is used for personal adornment, and has been ranked among the inferior gems. It is a decomposition of resinous vegetation, found with lignite or brown coal, of a lustrous velvet-black color, and capable of a beautiful polish, very soft and light, having a hardness of only one and one-half in the scale. It is the *gargates* of the ancients, named, says Pliny, from Garges, a place and river in Lycia, Asia Minor; or it may have been for Garges in Syria, since lignite is very abundant in that country. It is described as black, smooth, light, and combustible — qualities belonging to jet, and was employed for jewelry in ancient Rome, as it is at the present day in many countries. The early Britons had a predilection for jet, as is proved by numerous relics found in the Island of Great Britain. A set of jewelry made of this substance, and supposed to have belonged to a priestess of Cybele, was discovered in a stone coffin in one of the churches of Cologne, during repairs made in 1846.

Aude, in France, and Whitby, in England, are celebrated for their productions in this article. It is found at Whitby in the upper lias shale, and on the shores, where it is washed up by the sea after a storm, and collected by the natives with great labor. It occurs on the Baltic with amber, and is sometimes called black amber. It is a production of the United States, and some of the best specimens are said to come from Colorado.

It seems incredible that a material for jewelry so inexpensive and abundant should be counterfeited, yet great quantities of manufactured jet are sold in Spain and Turkey. There is evidence that many intagli cut in this substance and sold for antiques are recent forgeries.

CHAPTER XVIII.

THE QUARTZ FAMILY.

QUARTZ is a rock so plentifully distributed over the earth that its claims to be ranked as a precious stone seem presumptuous; but, as a fact, it yields a greater variety of gems of a second class than any other mineral species, and has been liberally used from time immemorial for engraving and jewelry. Its hardness renders it susceptible of a good polish, and protects it from the consequences of ordinary attrition; while its great diversity of colors, and low prices, are favorable for its general use in decoration.

The most common form of quartz crystal is hexagonal prisms, which range from transparent to opaque, and display nearly all the hues of the solar spectrum, diversified by numerous gradations. Quartz is infusible and insoluble, and possesses double refraction and electrical powers; its chemical constituent is pure silica, except in the colored varieties, which contain traces of certain other substances, as manganese, in amethyst, nickel, in chrysoprase, titania, in certain kinds of rose-quartz, and manganese and iron, in many red, green, yellow, and brown specimens.

The varieties are very numerous. Dana gives more than twenty used for art purposes; Church mentions a larger number, but makes, in some instances, two varieties where only slight differences exist, as cairngorm and smoky quartz, heliotrope and bloodstone. Westropp arranges the different kinds under three heads: first, Vitreous quartz, comprising crystal, amethyst, citrine, cairngorm, iris, rubasse, aventurine, and prase; second, Chalcedonyx, including carnelian, chalcedony, sardonyx, sard, onyx, nicolo, plasma, heliotrope, agate, mocha-stone (a variety of agate), cat's-eye, and chrysoprase; the third division, called Jasper-quartz, embraces bloodstone (differing from heliotrope), jasper, Egyptian pebbles, and porcellanite. King places aventurine with the jasper series.

Yellow quartz is sometimes called Scotch topaz, and Bohemian topaz, and the rose, Brazilian ruby, occasionally sold for spinel; the blue variety has frequently passed for water-sapphire, while specimens of brownish red are known as hyacinth of Compostella, and when impregnated with bituminous substances they are denominated smoky quartz, cairngorm, and Alençon diamonds.

Rock-crystal, or hyalin quartz, is the purest form of this mineral known among lapidaries, and is recognized by various names, as Bristol, Welsh, Irish, Cornish, and California diamonds, and sometimes it is employed in jewelry under the title of "white stone." It received the appellation of crystal ("ice"), from the nations of antiquity, who supposed it was formed by the excessive congelation of water, such as could be found only in the coldest regions. Claudianus, one of their writers, calls it "ice hardened into stone, which no frost could congeal nor dog-star dry up." Orpheus poetically calls the crystal "the translucent image of the Eternal Light," and suggests its use as a burning-glass to light the sacrificial flame. The East Indian believes it to be the mother or husk of the diamond.

The cavities frequently found in rock-crystals are sometimes filled with a fluid or gas which has given rise to some speculation; most scientists have thought it was water, but the experiments of Sir David Brewster led him to the opinion that the contents of these occupied cells consisted of an oleaginous substance. Whether the water or other matter was enclosed at the time of the formation of the crystal, or was afterwards infiltrated through its pores, is still a subject of dispute. When this mineral encloses very small fibres or slender prisms of rutile, oxide of titanium, or other substance, it is denominated Cupid's nets, Love's arrows, Venus' hair, and other fantastic names, and when it contains fibrous asbestos, it forms a variety called cat's-eye.

Vessels made of rock-crystal were highly valued by the Romans, for which they often paid enormous sums; for example, it is on record that a certain householder of ordinary means gave six thousand dollars for a crystal basin. The different museums of Europe comprise valuable collections of crystal cups, vases, and other articles, showing its extensive use in early times for such purposes; but few intagli of great age are found in this variety, though the Renaissance artists frequently employed it for engraving. It is shown by the inventory made in Paris in 1791 that the crown jewels comprised crystal goblets, vases, or other vessels, some of them being beautifully engraved, and were valued altogether at one million francs. One urn, measuring nine and one half inches in diameter, and nine in height, was engraved on the upper part with the figure of Noah asleep after his intoxication. The royal theatre of Sans Souci was lighted by a large chandelier made entirely of rock-crystal.

On account of a superstitious belief that crystal was incapable of holding poison, it became a favorite material for cups and goblets, especially to the ancient Romans. Nero is said to have possessed some magnificent cups of this kind, engraved with subjects from the Iliad; he is charged with breaking, in a fit of anger, two crystal goblets which cost nearly two thousand dollars.

Globes made of this variety of quartz were believed, in the Middle Ages, to have great magical powers, and were used for that purpose by Dr. Dee, a somewhat famous English astrologer of the sixteenth century. Rock-crystal is employed at the present day, not only for gem-stones, but also for lenses, polariscopes, and object-glasses for telescopes; it constitutes a very successful imitation of the diamond, and when artificially colored, passes for the ruby, sapphire, and other precious stones. The method of coloring is by plunging the heated crystal in a tincture of cochineal, or some other pigment, according to the hue desired, or by soaking it for some months in spirits of turpentine, saturated with some metallic oxide. The same result is secured by painting the back of the stone when set for jewelry. Some of the crystal intagli of the Renaissance are set with the engraved side downwards upon gold or azure foil, producing the effect of making the figures appear as if cut in relief upon the plain surface of a topaz or a sapphire. Rock-crystal artificially colored green, pink, and other tints, is often sold for beads in Switzerland and Germany.

Specimens of hyalin quartz of gigantic size are numerous, and may be seen in nearly all the museums of the world. The largest known to Pliny, which was consecrated by Julia Augusta as a sacred offering in the Capitol, weighed one hundred and fifty pounds; a specimen in the Jardin des Plantes, at Paris, weighs eight hundred pounds; a group in the museum of the University of Naples reaches nearly a ton; and another mass, mentioned by Dana, yields a weight of eight hundred and seventy pounds. Remarkable crystals from Brazil and Japan were exhibited at the French Exposition in 1866, and at

Philadelphia in 1876. Some of the mineral collections in the United States contain crystals of very large size.

Fine specimens found at Lake George and other places in New York, of great brilliancy, and sometimes with both ends of the crystal terminating in natural facets, are sold as "Lake George diamonds"; but even these "diamonds" are frequently counterfeited by glass and pastes. Quartz, occurring in the form of rolled pebbles, in New Jersey, Colorado, and Arkansas, affords a quality suitable for gems. The crystal employed for optical purposes in this country is almost entirely brought from Brazil — more on account of its cheapness, it has been said, than for any superiority over the native production.

Amethyst. — This name, when applied to a gem, is an epithet to denote color, and is given to varieties of other species, and sometimes to all stones of a purple hue. The term is from a word meaning "without intoxication," originally given to a kind of grape supposed to be free from any inebriating quality, and, by a figure of speech, applied to the substance holding the wine; hence it was considered the most suitable for drinking-cups, on account of its being a protection against intoxication. Its beautiful color, thought to be due to manganese, iron, and perhaps soda, can be dispelled by heat, a fact which has enabled lapidaries to secure uniformity of tint. The deep shades are less brilliant, and for this reason the artists of antiquity preferred the lighter colored specimens. Some antiquaries have believed the amethyst identical with the hyacinth of Pliny, while others have supposed the antique hyacinth was the same as modern sapphire. The oriental amethyst, a variety of the corundum, an exceedingly rare and beautiful gem, is distinguished from the quartz amethyst by its deep shade of violet without the reddish tint of the latter.

On account of its fine color, play of light, and capacity for polish, the amethyst once held a high rank as a decorative stone, perhaps next to the sapphire; but for the reason of the large importations from Brazil, its popularity declined in Europe, consequently its commercial value declined in corresponding ratio. As an illustration of this depreciation, it is stated, on competent authority, that a necklace of amethysts formerly valued at ten thousand dollars would not now command as many shillings. Most of the stones are cut in Germany, and appear to the best advantage as brilliants with a rounded table. They harmonize well with gold, diamonds, and pearls, but do not make an attractive evening ornament. It is the only one considered appropriate for mourning, and one of the gems most suitable for sacerdotal use.

Of all the quartz varieties the amethyst has been, probably, the most highly valued and the most frequently used for the art of engraving, in all periods, but ancient intagli, of all dates and in every style of work, occur almost invariably on the light-colored specimens, so that an engraving on a dark shade, says King, may be suspected as modern. Many Egyptian and Etruscan scarabei were on this variety of quartz. Among engraved amethysts of note was the gem bearing the likeness of the Emperor Trajan, which fell into the hands of Napoleon during his invasion of Prussia; a bust of Antonia, the daughter of Mark Antony; the head of a Syrian king, in the Pulsky collection; and an engraving by Dioscorides, in the National Library of Paris. Three superb oriental amethysts are mentioned in the inventory of the crown jewels of France, and a rare specimen, with bright red spots, or clouds, now belongs to the French collection.

Quartz amethysts are found in nearly every country, though Brazil, undoubtedly, yields the largest quantities, and like those from Siberia, they are frequently of gigantic size.

Very beautiful examples are known to occur in the environs of Carthagena, Spain; but India, and notably the Island of Ceylon, afford the best crystals. It is abundant in large masses in the United States, several localities furnishing excellent specimens for cabinet use.

Agate. — This stone, says Mr. Streeter, does not strictly belong to mineralogy, which deals with simple minerals. It is a conglomerate of certain quartz varieties, which, in color, texture, and translucency, are diverse, one from the other, as chalcedony, carnelian, jasper, and some others. When two or more of these precious stones form a cohesive mass and are arranged in stripes and spots, the combination is called agate. Some writers represent the agate as forming a group of gem-minerals, including nearly all the quartz varieties, a classification not generally adopted. In composition, it consists of ninety-eight per cent of silica; its different colors are supposed to be due to the presence of iron, manganese, bitumen, and chlorite. The beauty of the agate depends chiefly upon the character of the alternate layers of chalcedony and other varieties of quartz, of which it is composed, whether they are translucent, brilliant, of fine color, and capable of high polish.

Agates, in their native state, are frequently found in the cavities of igneous rocks caused by the escape of gas or steam, it is conjectured, when these rocks were in a fluid state. These cavities were subsequently filled with silica or some other mineral substance held in solution and deposited on the interior walls of these receptacles, thus often forming a kind of geode. These balls are sometimes furnished with a small funnel through which the silicious matter penetrated.

There are numerous varieties, dependent upon the arrangement of the layers or other incidental causes: as, when the stone presents delicate parallel lines of light and dark tints, it is called *banded*, or *ribbon* agate; when the colors are sharply defined, it is *onyx-agate;* when the stripes converge towards a centre, it is *eye-agate;* if divers colors are displayed, it becomes *iris* or *rainbow-agate;* when it has the semblance of moss enclosed, it is *mocha-stone* or *moss-agate;* and *dendritic*, when foliage and trees are simulated, though moss-agate and dendritic are generally classed as one variety. These vegetable representations are supposed to have been produced in water by some metallic particles, such as iron and manganese; they generally occur on a red, brown, or black ground of chalcedony or sapphirine. The name *mocha-stone* is derived from Mocha, in Arabia, where it was found. There are other varieties, which are known by the names of *jasper-agate*, *wood-agate* (wood petrified by agate), *undulata* or *zone-agate*, and *brecciated agate*. What are called Siena agates, seen in the Florence gallery, present a dark ground diversified with white, gray, brown, and yellow clouds.

The best specimens of agates are brought from India, and the second in quality are from Uruguay and Brazil; those found at Brighton, England, the Isle of Wight, Chamounix, and Niagara, are really the same in kind as the Indian and Brazilian, but differ from them by being water-worn. They are plentiful in the United States; one of the most noted localities for this gem is at Agate Bay, Lake Superior, where small, red, banded specimens are found, while large and beautiful agates have been discovered in the Rocky Mountains and in Colorado. The collection sent from that State to the New Orleans Exposition included many fine varieties. Of all American gems, the moss-agates are the most abundant and the cheapest; those found in streams, called "river-agates," are considered the best. They are nearly all sent abroad for cutting.

Oberstein and Idar, in Germany, are the chief centres of the agate industry. Those cut at these places were formerly obtained, to a great extent, from the hills bordering the River Nahe, but more recently they are brought from other countries, those from South America constituting a large part of the agates of commerce. The process, briefly described, of coloring this gem artificially, is as follows: The stone is first washed and dried, then laid in honey and water, and placed in a heated oven, care being taken that the water does not boil. This treatment is continued from fourteen to twenty-one days, after which the agate is removed from the honey and washed, then soaked in sulphuric acid, in a covered vessel overspread with hot ashes and burning coals. After being subjected to this roasting process a sufficient length of time, it is removed, washed, and dried. By this complicated treatment, the syrup in which it is immersed is absorbed by the more porous layers, then carbonized by the action of the acid, which forms the dark-colored varieties, but the red specimens are obtained by boiling the gems in a solution of protosulphate of iron, and then exposing them to heat. If laid in oil several hours, the agates acquire great brilliancy. The so-called Brazilian carnelians are prepared in a similar manner at the celebrated works of Oberstein. The jeweller of to-day, says Mr. King, can see no difference between the German silex, artificially stained, and the precious Indian agate, though the art student, and, it may be added, the practical mineralogist, are enabled at once to detect and appreciate the distinction.

This gem often exhibits some curious features of Nature's fantastic handiwork, as when it presents striking likenesses of certain persons, such as are seen in a specimen in the British Museum, with a portrait resembling the poet Chaucer; one in the Florence gallery, with a Cupid running; one in the Strawberry Hill collection, with a woman in profile; and another with the likeness of Voltaire. The most celebrated agate vase known was the two-handled cup engraved with bacchanalian subjects, and presented by Charles the Bald of France, in the ninth century, to the Abbey of St. Denis, and was used to hold the wine at the coronation of the French kings. This cup has been, by some mistake, regarded as the gift of Charles the Bold, Duke of Burgundy, of a later period. At the sack of Delhi, the English soldiers, with that "natural love of destruction," writes King, "which characterizes John Bull, smashed several chests of elegant agate cups."

The agate was prized as an ornamental stone by the Greeks and Romans, and has always been a favorite material for engraving; the Italo-Greek artists preferred the banded agate for this purpose. One of the largest and finest specimens of engraving on this gem represented the portrait of Alexander the Great. The ancient name of this variety of quartz was *achates*, from a river in Sicily where it was obtained. Pliny says the varieties were numerous, recognized by some distinctive feature, as wax-agate, smaragdus or green, blood, white, jasper, tree, undulated, coralline, and others.

Onyx. — This name is applied to a precious stone mentioned in the sacred writings, but it is not beyond doubt whether it might not have been a different gem from the modern onyx. The word signifies "finger-nail," given by the Greeks, who accounted for its origin in the following manner: While Venus was reposing in slumber on the banks of the Indus, Cupid, either from wanton sport or filial respect, cut her finger-nails with his arrows, and the parings, falling into the river, were converted into onyx. This myth indicates that the source of supply to the nations of antiquity was in oriental countries.

The onyx and the agate are similar, but unlike in the color

and arrangement of the layers: in the former, the different zones are parallel, while in the latter they are concentric, and sometimes the colors of the agate are disposed in the form of irregular clouds, veins, and spots, quite unlike those of the onyx. Alternations of light and dark chalcedony afford the specimens so much used in camei. The common variety of onyx has two opaque layers, of different colors, as black and white, dark red and white, green and white, and some other combinations, but the most frequent is blackish or brownish striped with white.

Some lapidaries consider the oriental variety superior to the occidental in several points: that it has a finer, closer texture; that it is harder, consequently receives a finer polish; that it is semi-transparent, and incapable of being artificially colored;—but Emanuel thinks there is no essential difference between the Indian and the German. The oriental onyx has three layers: the upper, red, blue, or brown; the middle, white or pearly; and the lower, black or brown. The common variety is frequently colored artificially, to imitate the Indian; the method of coloring is similar to that employed for agates, different chemical substances being added according to the tints required.

The onyx is occasionally found in so large masses that small pillars are cut from it; six of these occur in St. Peter's Church at Rome, and one is seen in the Temple of the Magi at Cologne. Pliny mentions thirty columns of large size in the banquet-hall of Callistus.

This gem was a favorite material with the ancients, who obtained it from Arabia and Persia, for cups and vases, fragments of which have been found in Roman remains. In their wars with Mithridates, the Romans carried off, as spoils of their victories, two thousand cups made of this stone, which had belonged to the vanquished king. One of the most celebrated antiques of the kind is the Mantuan Vase, seven inches high and two and one-half broad, cut from a single specimen, which offered to the lapidary a ground of a brown color for his reliefs of white and yellow figures representing Ceres and Triptolemus in search of Proserpine. Interesting examples of engraving on onyx are seen in the Vatican Library, at the Museum of Naples, at Venice, in the National Library of Paris, and in several other collections. The Dresden Museum contains a large specimen, measuring six and two-thirds inches by four and one-fourth, and the Austrian collection comprises one nine inches in diameter, thought to be the largest known among collectors.

Sardonyx. — This variety of quartz is a combination of the sard and the onyx, as the name implies. The Indian variety consists of a white, opaque layer superimposed upon a red, translucent zone of true sard; while the Arabian comprises no sard, but is formed of black and blue strata covered by one of opaque white, above which lies a third, of vermilion hue.

This gem was first worn at Rome by Africanus the Elder, when it became very fashionable, and, with the emerald, constituted the favorite ornamental stone of the Emperor Claudius. It has been employed for gem-engraving and for camei, and when used for the latter, the red layer forms the ground and the white band the figures; and if a third color, of milky white, occurs, as is sometimes the case, it serves for hair. A stone resembling the sardonyx, — it may be the eyed-onyx, or agate, — having a round spot in the centre, affords the jewellers an opportunity of introducing gold foil into a cavity made beneath, — an operation which imparts to the gem a remarkable brilliancy.

Nicolo. — This name, an abbreviation of *onicolo* ("little onyx") is applied both to a natural production and an artificial variety. It has been described as an onyx with a deep brown layer under a stratum of bluish white, so that the dark color underneath is seen through it. The Italian nicolo is represented as the onyx covered by bluish white spots encircled by milky zones, and by cutting out the spots an artificial kind is obtained. Nicolo is found in Bohemia and the Tyrol.

Chalcedony. — This mineral affords another illustration of the confusion introduced into the classification of precious stones, in consequence of the want of agreement among mineralogists. Westropp, with others, groups a large number of varieties under the head of chalcedonic quartz, and says pure chalcedony is colorless or pale horn color, but when tinted with iron or other substances, it forms an almost endless variety of sards, agates, carnelians, plasmas, etc. King defines it as a translucent variety of quartz, mixed with opal, and of a waxy appearance; Dana calls it a variety of quartz of a waxy lustre, transparent to translucent, varying in tints from white to black; while other writers represent chalcedony as a variety of quartz with alternate stripes of white and gray, as well as a term for all stones of the agate kind.

Many of the most beautiful silicious minerals are transparent chalcedonies of emerald green, purple, red, and blue tints. It is never found in regular crystals, and has very little lustre, but it is well adapted for engraving, for which it has been used from very remote times. A white chalcedony, with minute blood-red spots, has sometimes been called "St. Stephen's Stone."

Sapphirine is the name given by lapidaries to blue chalcedony, but this is distinct from the species of the same name, — a silicate of alumina, magnesia, and iron, of a pale blue or green color, with a hardness and specific gravity superior to those found in quartz.

Carnelian is thought to correspond to the *sardion* of Theophrastus and the *sarda* of Pliny, who says the name was derived from Sardis, where it was found. The etymology of carnelian is doubtful, having been referred to *caro* ("flesh"), to *cornu* ("horn"), to Caria (a province in Asia Minor), to an Arabic word signifying yellow, and to a Hebrew name for red. There is no doubt the term is an epithet denoting color, which varies from blood-red to wax-yellow or brown, and is capable of being intensified by heat. The blood-red holds the highest rank, and the pale red is next in value for ornamental stones. It receives a fine polish and forms one of the most desirable of the quartz gems; but, in consequence of large importations from Brazil, and the extensive business of artificial coloring, their commercial value has greatly depreciated, and their use for jewelry is much less than formerly, except in some countries, more especially in Germany and Poland, where they still hold an important rank.

The carnelian — *sarda* —was valued by the ancients more highly than the sapphire, and was used more generally for ornament than any other precious stone, and none, says Pliny, played so conspicuous a part in the comedies of Menander and Philemon. It was very generally employed in classic art for camei and engraving; the oldest of these remains are found in Germany. A carnelian with the portrait of Sextus Pompeius is in the Berlin collection; one engraved with Helen is in Vienna; another, with the head of Apollo, is in Florence; while St. Petersburg claims an antique on this gem, and the British Museum contains a specimen of fine workmanship in the form of a butterfly.

Sard has been identified with carnelian, and so closely resembles it as hardly to require a particular description; it may be classed as reddish brown and yellowish red varieties of chalcedony, the latter being styled oriental carnelian. Though Pliny derives the name from Sardis, others have referred it to *zerd*, a Persian word for yellow. The grades in tint seen in antique specimens are numerous. Sards from India were of very fine quality, but those from the neighborhood of Babylon were esteemed by the nations of antiquity of very great value.

Chrysoprase, "golden leek," is an opaque, apple-green chalcedony, colored by the oxide of nickel, and supposed to have been the prase of the ancients. It differs from plasma in vividness of tint, hardness, and opacity, and, though equal to the emerald in hue, it falls below it in lustre, and has the misfortune to lose its color when exposed to heat and sunshine, which, however, may be restored by immersion in a solution of nitrate of nickel.

This precious stone, on account of its capacity for polish and its agreeable color, once maintained an important position as a gem, and is still used for jewelry, to a considerable extent, on the continent of Europe It was frequently employed by the Greeks and Romans for intagli and camei, and is found in antique jewelry, particularly of Egyptian workmanship, set with lapis-lazuli. Existing in large masses, it has served the purposes of interior architectural decoration, as may be seen in the mosaic walls of St. Wenzel's Chapel, in the Cathedral of Prague, built in the fourteenth century. It was used by Frederick the Great to adorn Sans Souci, and the Old Palace at Potsdam ; in the latter building are seen two tables three feet long, two broad, and two inches thick, made of chrysoprase. The variety obtained from Silesia has long been celebrated as a decorative stone, and is frequently met with in that capacity.

Plasma is a translucent variety of chalcedonyx quartz, of bright green to leek-green, and resembles both prase and chrysoprase, yet constitutes a gem-stone distinct from both.

The ground color, due to some metallic oxide, is sometimes sprinkled with minute white or yellowish specks of some foreign substance. The name signifies "image," or anything formed or imitated. It is found in ancient ruins, but it was not used for ancient engraving, it is believed, before the later Roman Empire.

*Cat's-eye.** — This precious stone exhibits a peculiar opalescence, caused by fibres of asbestos running parallel across the stone. It is usually translucent, sometimes transparent, and displays red, blackish, yellowish green, and brown colors. This name has been given to different minerals presenting these peculiarities, constituting varieties of chrysoberyl, crocidolite, sapphire, and perhaps some others, as well as quartz. The chrysoberyl cat's-eye is a much more beautiful gem than the quartz variety, and is superior to it in hardness and specific gravity. The largest specimen of cat's-eye known belongs to this species, and was formerly owned by the King of Kandy, Ceylon, but is now in the collection of the South Kensington Museum.

Classic writers called this precious stone "wolf's-eye," and also *oculus belus*, because it was dedicated to the Assyrian god Belus.

Heliotrope, or Bloodstone. — There is no harmony between the names of this gem, and some writers make them distinct varieties. Westropp adópts this classification and calls heliotrope a translucent, green chalcedony or plasma, with blood-red spots, while the bloodstone he denominates a green jasper, interspersed with specks of a red color; but Dana says there is no essential difference between heliotrope and bloodstone.

* No less than ten different minerals afford the cat's-eye rays when cut in certain directions, so that the name is more appropriate to denote a particular method of cutting than a natural variety or species.

The name heliotrope was conferred upon this variety of quartz, from the idea that when plunged in water it presents a red reflection of the sun's image. There is a tradition, believed in the age of superstition, that at the Crucifixion the blood of Christ, falling upon a dark green jasper, produced the red spots. This stone was used as a talisman by the Gnostics and later Egyptians, and was in great demand in the Byzantine and Renaissance periods. The interesting myth that the heliotrope was invested with certain magical properties by incantation, when the person wearing it is made invisible, was boldly repudiated by Pliny, the practical old Roman, who had something of the iconoclastic tendencies of modern scientists.

Some Egyptian and Babylonian intagli are found on this variety, but engravings of this kind are rare, though they occur frequently with the Gnostics on green jasper mottled with brown.

Cairngorm, a name given to smoky quartz, is derived from a place in Scotland bearing the same title. Some authors designate the variety of intense color "morion," and the transparent brown or yellow crystals "cairngorm," but among lapidaries the cairngorm includes all shades of color, from black to yellow, known by different appellations, as Brazilian topaz, Mexican topaz, Spanish topaz, diamond of Alençon, smoke-stone, etc. It has been by some antiquaries identified with the *mormorion* of Pliny, who calls it a transparent stone of a deep black color,—a description better suited to the black tourmaline, as far as color is concerned, than to the cairngorm. This is a favorite gem with the Scotch jewellers, who cut it so as to display its color and brilliancy in a remarkable degree.

Aventurine.—This is an epithet applied to varieties of different species, having a peculiar construction, as well as to an artificial production, first made in Venice, by a mixture of powdered glass, protoxide of copper, and oxide of iron, heated for several hours. Quartz aventurine is a semi-transparent, brown, gray, reddish white or greenish white variety, interspersed with spangles of yellow mica, which glitter like gold, and, as it receives a high polish, it constitutes an attractive ornamental stone. It is found on the shores of the White Sea, in Siberia, Bohemia, Switzerland, France, Spain, and Scotland, while India produces a beautiful green variety.

Jasper.—The numerous varieties of this stone have given rise to different opinions about it, which renders a description and classification difficult. The word, derived from *iaspis*, has been rendered "green," "firm," or "tough." Pliny counts fourteen kinds of the *iaspis*, one being like crystal, which corresponds to the Scripture account of the jasper, one of the stones of the New Jerusalem, but does not answer to any variety known to us. The Indian green jasper of antiquity appears to have been a plasma of a rare kind, approaching the emerald in color, and it is possible their emerald was green jasper. The modern jasper is a compact variety of quartz, of various colors, comprising green, yellow, numerous shades of red, blue, and black, while among antique specimens are found vermilion and crimson hues. King considers the red jasper the bloodstone of the ancients.

The Egyptian jasper, characterized by intense red or ochre-yellow tints, deepening into chestnut-brown, sometimes spotted with black, was found near Cairo, and in the region of the Nile, and was extensively used in ancient art. What are known as Egyptian pebbles are composed of jasper, which frequently present an arborescent appearance. Red jasper was developed in Argos, Greece, and was a favorite with Roman engravers; while the Greeks preferred the yellow, also a native of their country, for artistic uses. A green, semi-transparent variety was more highly valued among the Romans, for engraving, than even the Greek red jasper, of which they made so free use. Vermilion jasper is seen only in antique work, but the source of supply is unknown to us; a white variety, resembling ivory, is said to be exceedingly rare.

European varieties display different greens variegated with other colors; that found in England is entirely green, or green spotted with red, flesh-color, or white. A kind called ribbon-jasper, or onyx-jasper, occurring in Saxony and Silesia, is made up of alternate bands or layers, usually red and green, sometimes purple and white; and when it unites a number of colors, it is known as "universal jasper." If agate and jasper are combined in the same specimens, they are agate-jasper or jasper-agate, according to the predominance of the one or the other. The Barga jasper, seen in the Florence Museum, is a very dark red, or reddish brown, and white stone, and the Corsican jasper, found in the same collection, exhibits rich green, purple, and gray tints. The "Pebbles of Rennes," mentioned by Haüy, are composed of agate and jasper of a very deep red ground interspersed with small round or oval spots of reddish or yellowish white, and were used for ornamental boxes and other similar work. The Arabian jasper is celebrated for the splendid dyes it affords. The red varieties are colored with peroxide of iron, the yellow and brown with hydrate of iron, and the green by chromate of iron.

The jasper was a favorite material for engraving and was very early used for that purpose, as we learn from the first breastplate made for the high priest of the Israelites, which contained an engraved jasper. Some fine portraits of the Roman emperors were cut on this precious stone, including a likeness of Nero, on a specimen weighing fifteen ounces. The head of Minerva on jasper, belonging to the Vatican collection, is considered the finest intaglio in existence; a high encomium, since there are so many examples of this kind of engraving of remarkable excellence. Jasper is used for the imperial seal of China, and has been employed in different countries for jewelry and ornamental articles of various kinds. In the Vatican Museum, there are two vases of remarkably beautiful jasper, one made of a bright red variety crossed with white veins, and the other black reticulated with fine yellow lines.

The poets have found this precious stone a suitable theme for their muse. Orpheus, in his "Lithika," says:—

"Full oft its hues the jasper's green displays,
The emerald's light, the blood-red sardion's blaze,
Sometimes vermilion, oft 'tis overspread
With the dull copper, or the apple-red."

Prase, Plasma, and *Chrysoprase* are all green varieties of quartz, and are liable to be confounded, though the shades of hue are different. The prase is of a dull leek-green color, and has little value as a gem-stone; the name, from *prason,* signifies "a leek," given in allusion to its hue.

Iris.—Jewellers give this name to a transparent, crystalline variety of quartz which reflects the hues of the rainbow, an effect due to the flaws and crevices in the interior of the stone. This peculiarity may be secured by an artificial process, either by dropping the crystals suddenly into boiling water, or by first heating them and then plunging them into cold water. Westropp mentions a variety with rose-colored seams, which he calls "rubasse," a substance generally produced by art. The iris was regarded with great favor in ancient and mediæval times, and even as late as Napoleon's day: the Empress Josephine possessed an elegant set of ornaments of remarkable "fire" and brilliancy, made of this variety of quartz.

Novaculite, a pure white quartz, developed at Hot Springs,

Arkansas, and known in that region as Ouachita ("oil stone"), is sometimes employed in jewelry.

As if the mineral kingdom had not furnished innumerable materials in the form of decorative and precious stones, vegetation has added a beautiful substance for the same purpose, in the nature of petrifactions, which rival the agate and other gems in the variety and richness of their hues, in the brilliancy of their lustre, and the hardness and compactness of their texture. Silicified woods of different kinds are found in Colorado, California, and other parts of the West, of suitable colors and hardness for ornaments, but none are comparable to those found in Arizona. A petrified forest has recently been discovered in this territory, covering an area of one thousand acres with prostrate trees, some of them more than one hundred and fifty feet in length, and from five to ten in diameter, which have been converted, by some mysterious process of nature, into trunks of precious stones, displaying all the prismatic colors in their varied combinations. Here we have representations of the amethyst, topaz, jasper, chalcedony, onyx, ruby, carbuncle, opal, malachite, agate, and others.

This silicified or opalized wood is of sufficient hardness to cut glass, and requires the diamond to work it, thus constituting one of the most desirable materials for fancy ornamental articles, mosaics, and jewelry. The Chalcedony Manufacturing Company, as it is called, has been organized for the purpose of utilizing this natural product and introducing it into the fine and useful arts, under the superintendence of Mr. William Adams, Jr., and Mr. George Stone, both of California, who have made several visits to the region of this remarkable "forest," in order to study its interesting features.

The species of trees which have undergone so wonderful a transformation, the time required for the process, and the agents employed, whether water holding in solution certain mineral substances, or gases, or some eruptive phenomenon, are questions which, probably, would divide scientists; but, whatever were the conditions necessary for such a result, the existence of these marvels cannot be questioned, and they are calculated to excite the wonder and admiration of every one who examines them. The exhibition of this marvellous production at the New Orleans Exposition formed one of its most interesting features, as it was one of the most novel. The exhibit included several *tons* of gems, in the condition of trunks of trees, varying in length and diameter, and combining a great variety of colors, representing a conglomerate of different precious stones, which were cut and polished under the eye of the spectator. The mosaics for tables and other ornamental articles afforded a remarkably fine selection for producing a beautiful harmony of colors.

APPENDIX A.

Size of the largest and the most remarkable diamonds known, as estimated by different writers, ranked as they at present exist, either cut or in their native state.

	Carats, uncut.	Carats, cut.
More than 1000 carats:—		
Braganza, or the King of Portugal's (its genuineness doubted)	1680 – 1880	
Between 300 and 400 carats:—		
Matan	367 – 387	
Nizam	340	
Between 200 and 300 carats:—		
Stewart.	288⅜	
Great Mogul (lost)	787½	279 9/16 – 280 – 297
Du Toit I..		244
Great Table		242½
Golconda	242	
Portuguese Regent		215 – 138½
Jagersfontein.	209¼	
Between 100 and 200 carats:—		
Orloff		193 – 194½
Darya-i-Nûr (Sea of Light).	232 (?)	186
Porter-Rhodes		150
Turkey I.		147
Tay-e-Mah (Crown of the Moon)		146 – 168
Austrian Yellow, or Florentine Brilliant . . .		139½
Abbas Murza, or Jehun Ghir Shah		130 – 138
Pitt, or Regent	410	136¾ – 137 – 139¾
Mountain of Splendor		135
Tiffany No. 1		125⅜
Star of the South	254½	125

	Carats, uncut.	Carats, cut.
Du Toit II.		124
Patrochino		120⅜
Moon of the Mountains		120
African Yellow		112
Star of Diamonds		107½
Cent Six		106
Koh-i-noor	$279\frac{9}{16}$	193 (1st cut) 102½ – 106½ (2d cut)
Rio-das-Velhas		105
Bazu		104
Raolconda		104 (nearly)
Hastings		101
Star of Beaufort		100
Between 75 and 100 carats:—		
Ahmedabad	157¼	94¼
Chapada		87½
Shah		93 – 86
Turkey II.		84
Throne		80 – 90
Nassack		89¾ (1st cut) 78⅝ (2d cut)
Tiffany No. 2 (canary color)		77
English Dresden	119½	76½
Between 25 and 75 carats:—		
Akbah Shah	116	71 – 72
Shah Jehan	unknown	
Star of Sarawak		68 – 70
Russian Table		68
Mascarenhas I.		67½
Sea of Glory		66
Tennant	112	66
Couloui		63⅜ – 49½
Mascarenhas II.		57
Ascot Brilliants		56
Savoy		54
Pear		53¾ – 54¾
Great Sancy		53¼ – 54
Tavernier A. B. C.		$51\frac{9}{16}$ – 32⅜ – 31⅜
Eugénie		51
Queen of the Belgians		50
Banian		48½
Bavarian		48½

	Carats, uncut.	Carats, cut.
Three Tables of Tavernier		48½ to 52½
Pigott		47½ – 82½
Antwerp		47½
Star of South Africa, or Dudley	83½	46½
Tavernier, or French Blue	112¼	67 (1st cut)
		44¼ (2d cut)
Hope Blue		44¼
Cleveland		42½
Dresden Green		40½ – 48½
Polar Star		40
Green Brilliant		40
Pasha of Egypt		40
Dresden White I. and II.		39+ – 30¾
Holland		36
Bantam		36
Hornby		36
Munich Blue		36
Heart		35
Napoleon		34 – 35
Little Sancy		34
Cumberland		32
Brazilian	90	32
Crown		31⅝ – 32
Dresden Yellow (four diamonds)		30 each
Below 25 carats:—		
Halphen (rose-colored)		22½
Paul First's Diamond (ruby-red)		10
Weight unknown: Charles the Bold's Diamond; Sun of the Sea; Sea of Light.		

APPENDIX B.

CLASSIFICATION OF PRECIOUS STONES ACCORDING TO THEIR PRINCIPAL CONSTITUENTS.

Carbon (pure)	Diamond.
Alumina (nearly pure)	Precious corundum (or sapphire), ruby, oriental emerald, oriental topaz, oriental amethyst, oriental aquamarine, asteria (or star-ruby), star-sapphire, girasol.
Alumina, water	Diaspore.
Alumina, magnesia, iron . . .	Spinel, balas, rubicelle, almandine, ceylonite.
Alumina, glucina	Chrysoberyl (or oriental chrysolite), cymophane, alexandrite, cat's-eye.
Silica, alumina, glucina	Beryl, emerald (occidental), aquamarine.
Silica, alumina, magnesia, iron, lime, manganese	Garnet, carbuncle, almandite (or almandine), pyrope, essonite (or cinnamon-stone), seriam, uwarowite, asteria (or star-garnet), vermeille (or hyacinth-garnet), bobrowska (not beyond doubt), grossularite, melanite.
Silica, alumina, fluorine . . .	Topaz (occidental), Brazilian sapphire (or blue topaz), gouttes d'eau.
Silica, magnesia, iron	Chrysolite, peridot, olivine.
Silica, alumina, lime, magnesia .	Vesuvianite, or idocrase, hyacinth, cyprine.
Silica, alumina, magnesia, iron, lime	Hypersthene.
Silica, lime, magnesia, iron . .	Diopside (pyroxene).
Silica, alumina, magnesia, iron .	Iolite, or dichroite (sometimes called sapphire d'eau).
Silica, alumina, glucina, water .	Euclase.
Silica, alumina, iron	Chiastolite (macle).
Silica, zinc	Willemite.

Silica, alumina, magnesia, lime, iron, water	Jade, or nephrite.
Silica, alumina, water, potash, etc.	Agalmatolite.
Silica, alumina, lime, iron, manganese	Axinite.
Silica, lime, magnesia, iron	Crocidolite, called also blue asbestos.
Silica, water	Opal, hydrophane, hyalite, cachelong, girasol, moss-opal.
Silica, copper, water	Dioptase.
Silica, alumina, magnesia, boron, fluorine, soda, potash, iron, lime, etc.	Tourmaline, rubellite, indicolite, achroite, aphrizite, peridot of Brazil, peridot of Ceylon, Brazilian emerald, Brazilian ruby, Brazilian sapphire.
Silica, zirconia	Zircon, hyacinth, jacinth, jargoon.
Silica, alumina, potash, soda, lime, etc.	Moonstone (orthoclase), sunstone (oligoclase), Amazon-stone, aventurine, adularia, elaeolite (nephelite).
Silica, soda, lime, sulphur	Lapis-lazuli.
Silica, magnesia, fluorine, iron	Chondrodite.
Silica, alumina, lime, soda, potash, iron	Labradorite.
Silica, alumina	Kyanite.
Silica, alumina, lithia	Spodumene, hiddenite (or lithia-emerald).
Silica, alumina, iron, lime, water, soda	Chlorastrolite.
Silica, alumina, lime, iron, etc.	Epidote.
Silica, alumina, potash, lithia, etc.	Lepidolite.
Silica, glucina	Phenakite.
Silica, alumina, lime, soda, iron, water	Zonochlorite.
Silica, magnesia, iron, water	Serpentine, bowenite, williamsite.
Silica, alumina, potash	Obsidian (orthoclase).
Silica, alumina, lime, soda, water,	Thomsonite.
Silica, magnesia, iron, water	Steatite.
Silica, magnesia	Rhodonite.
Silica, lime, potash, some fluorine	Apophyllite.

Components	Stones
Silica, titanic acid, lime . . .	Titanite.
Silica (pure)*	Quartz, rock-crystal, amethyst, sapphirine-quartz, citrine, cairngorm, iris, rubasse, aventurine, prase, chalcedony, carnelian, sard, sardonyx, onyx, nicolo, plasma, agate, heliotrope (or bloodstone), mocha-stone, cat's-eye, chrysoprase, novaculite, sagenite, jasper, porcellanite, Egyptian pebbles, Scotch topaz, Bohemian topaz, Brazilian topaz, Mexican topaz, hyacinth of Compostella, Alençon diamonds, Bristol, Welsh, Irish, Cornish, Hot Springs, and California diamonds.
Arsenic acid, copper, water . .	Euchroite.
Phosphoric acid, alumina, copper, iron	Turquoise, odontolite (bone colored by copper).
Phosphoric acid, cerium, lanthanum, thoria, manganese . . .	Monazite.
Sulphuric acid, lime, water . .	Selenite.
Sulphuric acid, baryta	Barite.
Carbonic acid, manganese . . .	Rhodochrosite.
Carbonic acid, lime	Calcite, pearl with organic matter.
Chromic acid, lead	Crocoite, or crocoisite.
Fluorine, lime	Fluorite.
Titanic acid, iron	Rutile, octahedrite.
Lime, sulphuric acid	Vulpinite.
Metallic minerals, iron-bearing .	Magnetite, hematite, marcasite, pyrite.
Metallic minerals, copper-bearing	Malachite, azurite, dioptase.
Carbon, hydrogen, oxygen . .	Succinite, or amber.

* Some foreign substances in colored varieties.

APPENDIX C.

HARDNESS AND SPECIFIC GRAVITY OF PRECIOUS STONES.

	Hardness.	Specific gravity.		Hardness.	Specific gravity.
Diamond	10	3.52	Rubellite		
Corundum varieties .	9	3.9+	Indicolite		
Oriental emerald . .			Achroite		
amethyst .			Aphrizite		
topaz . . .			Peridot of Ceylon .		
aquamarine.			Peridot of Brazil . .		
peridot . .			Brazilian emerald .		
Girasol			Brazilian ruby . . .		
Asteria			Iolite, or dichroite . .	7 - 7.5	2.5 - 2.6
Sapphire		4	Hiddenite (spodu-		
Ruby	8.8		mene)	7 - 7.5	3.19
Chrysoberyl varieties .	8.5	3.5 - 3.8	Quartz varieties . .	7	2.5 - 2.8
Cymophane . . .			Agate		
Alexandrite			Agate jasper . . .		
Cat's-eye			Amethyst		
Topaz (occidental) .	8	3.4 - 3.65	Aventurine		
Brazilian sapphire, or			Bloodstone		
blue topaz . . .			Cairngorm		
Spinel varieties . . .	8	3.5 - 4.1	Cat's-eye		
Balas, ceylonite . .			Chalcedony . . .		
Rubicelle			Chrysoprase . . .		
Almandine			Carnelian		
Beryl varieties . . .	7.5-8	2.6 - 2.7	Egyptian jasper . .		
Aquamarine . . .			Heliotrope		
Emerald (occidental)			Milky quartz . . .		
Phenakite	7.5-8	2.96-3	Novaculite		
Zircon varieties . . .	7.5-7.8	4.05-4.75	Onyx		
Hyacinth, jacinth, or			Plasma		
jargoon			Porcelain-jasper . .		
Andalusite	7.5	3 - 3.5	Prase		
Euclase	7.5	3.09	Ribbon-jasper . . .		
Tourmaline varieties .	7 - 7.5	2.9 - 3.3	Rock-crystal . . .		

	Hardness.	Specific gravity.		Hardness.	Specific gravity.
Rose-quartz			Labradorite	6	2.6 – 2.7
Sapphirine			Opal varieties	5.5–6.5	1.9 – 2.3
Sard			Cachelong		
Sardonyx			Hydrophane		
Sagenite			Hyalite		
Scotch topaz			Magnetite, or lode-		
Garnet varieties	6.8–7.9	4.1 – 4.3	stone	5.5–6.5	4.9 – 5.2
Almandine, or alman-			Hematite	5.5–6.5	4.5 – 5.3
dite			Chlorastrolite	5.5–6	3.18
Asteria			Bowenite (serpentine)	5.5–6	2.5 – 2.7
Carbuncle			Willemite	5.5	3.89–4.18
Essonite, or cinna-			Octahedrite	5.5	3.8 – 3.9
mon-stone			Kyanite	5 – 7.2	3.4 – 3.7
Hyacinth-garnet, or			Diopside (pyroxene)	5 – 6	3.2 – 3.5
vermeille			Hypersthene	5 – 6	3 – 3.9
Green garnet			Lapis-lazuli	5 – 5.5	2.3 – 2.4
Uwarowite			Thomsonite	5 – 5.5	2.3 – 2.4
Pyrope			Monazite	5 – 5.5	4.9 – 5.2
Seriam			Titanite	5 – 5.5	3.4 – 3.5
Epidote	6.7	3.2 – 3.5	Dioptase	5	3.2 – 3.3
Zonochlorite	6.5–7.2	3.04–3.15	Apophyllite	4.5–5	2.3 – 2.4
Jadeite	6.5–7	3.3	Williamsite (serpen-		
Axinite	6.5–7	3.27	tine)	4.5	2.5 – 2.6
Spodumene	6.5–7	3.13–3.19	Crocidolite	4	3.2
Diaspore	6.5–7	3.3 – 3.5	Rodochrosite	3.5–4.5	3.4 – 3.7
Vesuvianite, or ido-			Azurite	3.5–4.2	3.5 – 3.8
crase	6.5	3.3 – 3.4	Euchroite	3.5–4	3.38
Chrysolite varieties	6.3–7	3.3 – 3.5	Malachite	3.5–4	3.7 – 4.01
Peridot			Vulpinite (anhydrite)	3–3.5	2.8 – 2.9
Olivine			Noble serpentine	2.5–5	2.5 – 2.6
Moonstone (ortho-			Lepidolite	2.5–4	2.8 – 3
clase)	6 – 7	2.5 – 2.6	Barite	2.5–3.5	4.3 – 4.7
Sunstone (oligoclase)	6 – 7	2.5 – 2.7	Agalmatolite (pinite)	2.5–3.5	2.6 – 2.8
Adularia	6 – 6.5	2.4 – 2.6	Calcite	2.5–3.5	2.5 – 2.7
Pyrite	6 – 6.5	4.8 – 5.2	Crocoite	2.5–3	5.9 – 6.1
Marcasite	6 – 6.5	4.6 – 4.8	Amber (succinite)	2–2.5	1.06
Rutile	6 – 6.5	4.18–4.25	Gypsum (selenite)	1.5–2	2.3
Amazon-stone (ortho-			Steatite (talc)	1–1.5	2.5 – 2.8
clase	6 – 6.5	2.4 – 2.6	Jet (coal)	0.5–2.5	1 – 1.8
Chondrodite	6 – 6.5	3.1 – 3.2	Gems of animal ori-		
Jade (nephrite)	6 – 6.5	2.9 – 3.1	gin:—		
Obsidian (orthoclase)	6	2.2 – 2.8	Pearl	2.5–3.5	2.5 – 2.7
Turquoise, or callaite	6	2.6 – 2.8	Coral	3	

APPENDIX D.

RELATIVE HARDNESS OF PRECIOUS STONES. — SCALE, FROM 1 TO 10.

First rank, 10 . . .	Diamond.
Second rank, 9 . . .	Varieties of the precious corundum (Ruby 8.8).
Third rank, 8+ . . .	Chrysoberyl, spinel, topaz.
Fourth rank, 7+ . .	Andalusite, beryl, euclase, hiddenite, iolite, phenakite, quartz, tourmaline, zircon.
Fifth rank, 6+ . . .	Amazon-stone (or microcline), adularia, axinite, chondrodite, chrysolite, diaspore, epidote, garnet, jade, jadeite, labradorite, marcasite, moonstone, obsidian, pyrite, rutile, spodumene, sunstone, turquoise, vesuvianite (or idocrase), zonochlorite.
Sixth rank, 5+ . . .	Bowenite, chlorastrolite, diopside, dioptase, hematite, hypersthene, kyanite, lapis-lazuli, magnetite (or lodestone), monazite, octahedrite, opal, thomsonite, titanite, willemite.
Seventh rank, 4+ . .	Crocidolite, williamsite, apophyllite.
Eighth rank, 3+ . .	Azurite, euchroite, malachite, rhodocrosite, vulpinite.
Ninth rank, 2+ . . .	Agalmatolite, amber, barite, calcite, crocoite, lepidolite, serpentine.
Tenth rank, 1+ . . .	Jet, selenite, steatite.

APPENDIX E.

RELATIVE SPECIFIC GRAVITY OF PRECIOUS STONES.

First rank, 5+ . . . Crocoite.

Second rank, 4+ . . Barite, hematite, magnetite, marcasite, monazite, pyrite, rutile, zircon.

Third rank, 3+ . . . Axinite, azurite, chlorastrolite, chondrodite, chiastolite (andalusite), crocidolite, chrysolite, chrysoberyl, corundum, diamond, diaspore, diopside, dioptase, epidote, euchroite, euclase, garnet, hiddenite, hypersthene, jadeite, kyanite, malachite, octahedrite, rhodocrosite, spinel, spodumene, titanite, topaz, vesuvianite (or idocrase), willemite, zonochlorite.

Fourth rank, 2+ . . Adularia, agalmatolite, amazon-stone, apophyllite, beryl, bowenite, calcite, iolite (or dichroite), jade (or nephrite), labradorite, lapis-lazuli, lepidolite, moonstone, obsidian, phenakite, quartz, serpentine, selenite, steatite, sunstone, thomsonite, tourmaline, turquoise, vulpinite, williamsite.

Fifth rank, 1+ . . . Amber, jet, opal.

APPENDIX F.

LOCALITIES IN THE UNITED STATES WHERE GEM-MINERALS HAVE BEEN FOUND.

Adularia	Maine, Massachusetts, Connecticut.
Agalmatolite	North Carolina.
Agate	Massachusetts, Rhode Island, Connecticut, North Carolina, Colorado, Texas, Rocky Mountains, California, Oregon, Arizona, Lake Superior, Yellowstone Park.
Agatized wood . . .	Arizona, Colorado.
Amazon-stone . . .	New York, Colorado.
Amber	Massachusetts, New York, New Jersey, Maryland, Delaware.
Amethyst (quartz) . .	Maine, New Hampshire, Rhode Island, Pennsylvania, Virginia, North Carolina, Georgia, Michigan, Texas, Colorado, and other places.
Andalusite	Maine, Massachusetts, New Hampshire, Vermont, Connecticut, Pennsylvania, California.
Anhydrite	New York.
Apatite	Maine, New Hampshire, Massachusetts, Connecticut, New York, New Jersey, Pennsylvania, Delaware, Maryland, Kentucky.
Apophyllite	Maine, New Jersey, Lake Superior.
Aquamarine	Maine, New Hampshire, Vermont, Massachusetts, Pennsylvania, North Carolina.
Axinite	Maine, New York, Pennsylvania, Wisconsin.
Azurite	New Hampshire, New York, Pennsylvania, New Jersey, Wisconsin, North Carolina, Tennessee, Missouri, Arkansas, California.
Barite	Found in most of the States and Territories.
Beryl	Maine, New Hampshire, Massachusetts, Connecticut, Pennsylvania, North Carolina.
Bowenite	Rhode Island, Pennsylvania.

Cairngorm	Maine, New Hampshire, Vermont, Massachusetts, Connecticut, Pennsylvania, Colorado.
Chalcedony	Rhode Island, Texas, Colorado, Rocky Mountains.
Chiastolite (andalusite),	Maine, New Hampshire, Vermont, Massachusetts, Connecticut, Pennsylvania, California.
Chlorastrolite . . .	Lake Superior.
Chondrodite	Massachusetts, New York, New Jersey, Pennsylvania.
Chrysoberyl	Maine, New Hampshire, Vermont, Connecticut, New York.
Chrysolite.	Arizona, Montana, New Mexico, Colorado, Vermont.
Corundum	Found in several of the States and Territories.
Crocoisite	California.
Diamond	Virginia, North Carolina, Georgia, Indiana, Wisconsin, Oregon, Colorado, California. [Generally in isolated crystals.]
Diaspore	Massachusetts, Pennsylvania, North Carolina.
Diopside (pyroxene) .	New York, Arizona, New Mexico.
Elaeolite (nephelite) .	Arkansas, Massachusetts, Maine.
Emerald	North Carolina.
Epidote	New Hampshire, Massachusetts, Connecticut, Rhode Island, New York, New Jersey, Pennsylvania, Michigan.
Garnet	Maine, New Hampshire, Vermont, Massachusetts, Connecticut, New York, New Jersey, Delaware, Pennsylvania, Wisconsin, North Carolina, Georgia, Texas, California, Colorado, Arizona, New Mexico, Montana, Alaska.
Hematite	Found in many of the States and Territories.
Hiddenite (spodumene)	North Carolina.
Idocrase (vesuvianite)	Maine, New Hampshire, Massachusetts, New York, New Jersey, Colorado.
Iolite (or dichroite) .	New Hampshire, Massachusetts, Connecticut.
Jadeite	Pennsylvania, Alaska.
Jasper	Massachusetts, New York, Kansas, Colorado, Texas, and other places.
Jet	Texas, Colorado.
Kyanite	New Hampshire, Vermont, Massachusetts, Connecticut, Pennsylvania, Maryland, Virginia, North Carolina.
Labradorite	New York, New Jersey, Pennsylvania, Arkansas.
Lepidolite.	Maine, Massachusetts, Connecticut.

Magnetite	Found in most of the States and Territories.
Malachite	Maine, New Hampshire, Vermont, Connecticut, New Jersey, Pennsylvania, Wisconsin, Maryland, Virginia, North Carolina, South Carolina, Tennessee, Missouri, Arkansas, Arizona, California.
Monazite	Connecticut, New York, North Carolina.
Moonstone (orthoclase)	Pennsylvania, Virginia.
Moss-agate	North Carolina, Colorado, Montana, Wyoming.
Novaculite	North Carolina, Georgia, Arkansas.
Obsidian	California, Colorado, New Mexico, Nevada.
Octahedrite	Rhode Island.
Olivine	Vermont, New Mexico, Montana, Arizona.
Opal	New York, Pennsylvania, North Carolina, Georgia, Florida, Colorado, Arizona, California.
Peridot	Vermont, New Mexico, Colorado, Montana, Arizona.
Phenakite	Colorado.
Prehnite	Massachusetts, Connecticut, New York, New Jersey, Lake Superior.
Pyrite	Found in numerous localities.
Quartz varieties	Found in numerous localities.
Rhodonite	Maine, Massachusetts, Vermont, New Hampshire, Rhode Island, New Jersey, Montana.
Rhodocrosite	New York, New Jersey, Nevada.
Ruby (corundum)	New Jersey, Pennsylvania, Virginia, North Carolina, Georgia, Colorado, New Mexico, Montana, Arizona.
Rutile	Maine, New Hampshire, Vermont, Massachusetts, Connecticut, New York, New Jersey, Pennsylvania, North Carolina, Georgia, Arkansas.
Sagenite (or Venus'-hair stone)	North Carolina.
Sapphire (corundum)	New Jersey, Pennsylvania, Virginia, North Carolina, Georgia, Colorado, New Mexico, Montana, Arizona.
Selenite (gypsum)	Found in numerous places.
Spinel	New York, New Jersey, North Carolina, Georgia, Colorado.
Spodumene	Maine, New Hampshire, Massachusetts, Connecticut, North Carolina.
Steatite (talc)	Found in many of the States.
Sunstone	Connecticut, New York, Pennsylvania.

Thetis'-hair stone . .	Rhode Island.
Thomsonite	Lake Superior, Arkansas (in the Ozark Mountains).
Titanite	Maine, Massachusetts, Connecticut, New York, New Jersey, Pennsylvania.
Topaz	Maine, Connecticut, Arkansas, Colorado, Arizona, New Mexico, Utah.
Tourmaline	Maine, New Hampshire, Vermont, Massachusetts, Connecticut, New York, New Jersey, Pennsylvania, South Carolina, Texas, California.
Turquoise	Arizona, New Mexico, Nevada.
Vulpinite	New York.
Willemite	New Jersey.
Williamsite	Pennsylvania.
Zircon	Maine, Vermont, Connecticut, New York, New Jersey, Pennsylvania, North Carolina, California.
Zonochlorite	Lake Superior.
Corals (fossil) . . .	Iowa.
Pearls	California, Texas, Ohio, Tennessee.

INDEX.

www.ingramcontent.com/pod-product-compliance
Lightning Source LLC
Chambersburg PA
CBHW021352210326
41599CB00011B/847

* 9 7 8 3 3 3 7 0 2 4 8 5 7 *